NANOTECHNOLOGY SCIENCE AND TECHNOLOGY

ELECTROSPINNING OF NANOFIBERS

FROM INTRODUCTION TO APPLICATION

NANOTECHNOLOGY SCIENCE AND TECHNOLOGY

Additional books in this series can be found on Nova's website under the Series tab.

Additional E-books in this series can be found on Nova's website under the E-books tab.

NANOTECHNOLOGY SCIENCE AND TECHNOLOGY

ELECTROSPINNING OF NANOFIBERS

FROM INTRODUCTION TO APPLICATION

A. K. HAGHI

EDITED BY
G. E. ZAIKOV

Nova Science Publishers, Inc.
New York

For permission to use material from this book please contact us:
Telephone 631-231-7269; Fax 631-231-8175
Web Site: http://www.novapublishers.com

Additional color graphics may be available in the e-book version of this book.

LIBRARY OF CONGRESS CATALOGING-IN-PUBLICATION DATA

Electrospinning of nanofibers from introduction to application / authors,
A.K. Haghi, G.E. Zaikov.
p. cm.
Includes bibliographical references and index.
ISBN 978-1-61122-865-6 (hardcover)
1. Nanofibers. 2. Textile fibers, Synthetic. 3. Electrospinning. I.
Haghi, A. K. II. Zaikov, G. E. (Gennadii Efremovich), 1935-
TA418.9.F5E4385 2010
620'.5--dc22

2010043906

Published by Nova Science Publishers, Inc. † New York

This volume is dedicated to the memory of Frank Columbus

On December 1st 2010, Frank H. Columbus Jr. (President and Editor-in-Chief of Nova Science Publishers, New York) passed away suddenly at his home in New York.

We lost our colleague, our good friend, a nearly perfect person who helped scientists from all over the world. Particularly Frank did much for the popularization of Russian and Georgian scientific research, publishing a few thousand books based on the research of Soviet (Russian, Georgian, Ukranian etc.) scientists.

Frank was born on February 26th 1941 in Pennsylvania. He joined the army upon graduation of high school and went on to complete his education at the University of Maryland and at George Washington University. In 1969, he became the Vice-President of Cambridge Scientific. In 1975, he was invited to work for Plenum Publishing where he was the Vice-President until 1985, when he founded Nova Science Publishers, Inc.

Frank Columbus did a lot for the prosperity of many Soviet (Russian, Georgian, Ukranian, Armenian, Kazakh, Kyrgiz, etc.) scientists publishing books with achievements of their research. He did the same for scientists from East Europe – Poland, Hungary, Czeckoslovakia (today it is Czeck republic and Slovakia), Romania and Bulgaria.

He was a unique person who enjoyed studying throughout the course of his life, who felt at home in his country which he loved and was proud of, as well as in Russia and Georgia.

There is a famous Russian proverb: "The man is alive if people remember him." In this case, Frank is alive and will always be in our memories while we are living. He will be remembered for his talent, professionalism, brilliant ideas and above all – for his heart.

CONTENTS

PREFACE

Nanotechnology is revolutionizing the world of materials. The research and development of nanofibers has gained much prominence in recent years due to the heightened awareness of its potential applications in the medical, engineering and defense fields. Among the most successful methods for producing nanofibers is the electrospinning process. Electrospinning introduces a new level of versatility and broader range of materials into the micro/nanofiber range. An old technology, electrospinning has been rediscovered, refined, and expanded into non-textile applications.

Electrospinning has the unique ability to produce ultrathin fibers from a rich variety of materials that include polymers, inorganic or organic compounds and blends. With the enormous increase of research interest in electrospun nanofibers, there is a strong need for a comprehensive review of electrospinning in a systematic fashion. With the emergence of nanotechnology, researchers become more interested in studying the unique properties of nanoscale materials. Electrospinning, an electrostatic fiber fabrication technique has evinced more interest and attention in recent years due to its versatility and potential for applications in diverse fields. The notable applications include in tissue engineering, biosensors, filtration, wound dressings, drug delivery, and enzyme immobilization. The nanoscale fibers are generated by the application of strong electric field on polymer solution or melt. The non-wovens nanofibrous mats produced by this technique mimics extracellular matrix components much closely as compared to the conventional techniques. The sub-micron range spun fibers produced by this process, offer various advantages like high surface area to volume ratio, tunable porosity and the ability to manipulate nanofiber composition in order to get desired properties and function. Over the years, more than 200 polymers have been electrospun for various applications and the number is still increasing gradually with time.

The discovery and rapid evolution of electrospinning have led to a vastly improved understanding of nanotechnology, as well as dozens of possible applications for nanomaterials of different shapes and sizes ranging from composites to biology, medicine, energy, transportation, and electronic devices.

Electrospinning is a highly versatile method to process solutions or melts, mainly of polymers, into continuous fibers with diameters ranging from a few micrometers to a few nanometers. This technique is applicable to virtually every soluble or fusible polymer. The polymers can be chemically modified and can also be tailored with additives ranging from simple carbon-black particles to complex species such as enzymes, viruses, and bacteria.

Electrospinning appears to be straightforward, but is a rather intricate process that depends on a multitude of molecular, process, and technical parameters. The method provides access to entirely new materials, which may have complex chemical structures. Electrospinning is not only a focus of intense academic investigation; the technique is already being applied in many technological areas.

This book presents some fascinating phenomena associated with the remarkable features of nanofibers in electrospinning processes and new progress in applications of electrospun nanofibers.

This new book offers an overview of structure–property relationships, synthesis and purification, and potential applications of electrospun nanofibers. The collection of topics in this book aims to reflect the diversity of recent advances in *electrospun nanofibers* with a broad perspective which may be useful for scientists as well as for graduate students and engineers.

Chapter 1

MECHANISM OF ELECTROSPINNING PROCESS

INTRODUCTION

Electrospinning is a novel and efficient method by which fibers with diameters in nanometer scale entitled as nanofibers, can be achieved. In electrospinning process, a strong electric field is applied on a droplet of polymer solution (or melt) held by its surface tension at the tip of a syringe's needle (or a capillary tube). As a result, the pendent drop will become highly electrified and the induced charges are distributed over its surface. Increasing the intensity of electric field, the surface of the liquid drop will be distorted to a conical shape known as the Taylor cone [1]. Once the electric field strength exceeds a threshold value, the repulsive electric force dominates the surface tension of the liquid and a stable jet emerges from the cone tip. The charged jet is then accelerated toward the target and rapidly thins and dries as a result of elongation and solvent evaporation. As the jet diameter decreases, the surface charge density increases and the resulting high repulsive forces split the jet to smaller jets. This phenomenon may take place several times leading to many small jets. Ultimately, solidification is carried out and fibers are deposited on the surface of the collector as a randomly oriented nonwoven mat [2]-[5]. Figure 1 shows a schematic illustration of electrospinning setup.

The objective of this paper is to use RSM to establish quantitative relationships between electrospinning parameters and mean and standard deviation of fiber diameter as well as to evaluate the effectiveness of the empirical models with a set of test data.

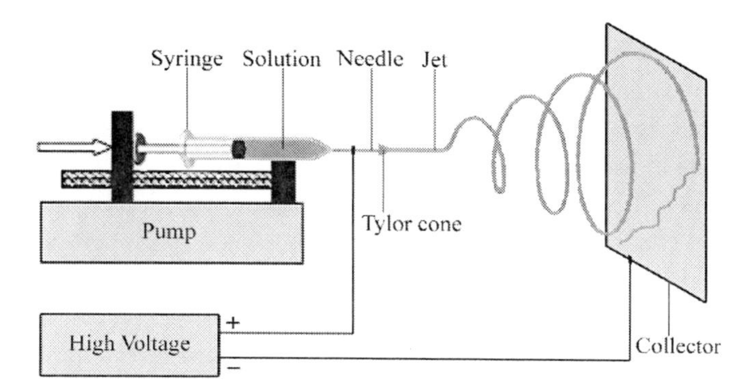

Figure 1. A typical image of Electrospinning process [6].

EXPERIMENTAL

Commercial Polyacrylonitrile (PAN) polymer containing 6% mathtlacrylate with molecular weight (Mw) of 100000 was supplied by Merk. N-methyl-2- pyrolidon (NMP) was from Riedel-de Haën. Aniline from Merck was vacuum distilled prior to use. The Polyaniline (PANI) used was synthesized in our laboratory.

Polyaniline was synthesized by the oxidative polymerization of aniline in acidic media. 3 ml of distilled aniline was dissolved in 150 ml of 1N HCl and kept at 0-5 °C. 7.325g of $(NH_4)_2S_2O_8$ was dissolved in 35 ml of 1N HCl and added drop wise under constant stirring to the aniline/HCl solution over a period of 20 minutes. The resulting dark green solution was maintained under constant stirring for 4 hrs. The prepared suspension was dialyzed in a cellulose tubular membrane (Dialysis Tubing D9527, molecular cutoff = 12,400, Sigma) against distilled water for 48 hours. Then it was filtered and washed with water and methanol. The synthesized Polyaniline was added to 150 mL of 1N (NH4) OH solution. After an additional 4 hrs the solution was filtered and a deep blue emeraldine base form of Polyaniline was obtained (PANIEB).The synthesized Polyaniline was dried and crushed into fine powder and then passed trough a 100 mesh. Intrinsic viscosity of the synthesized Polyaniline dissolved at sulfuric acid (98%) was 1.18 dl/g at 25 °C.

The PANI solution with concentration of 5 %(W/W) was prepared by dissolving exact amount of PANI in NMP. The PANI was slowly added to the NMP with constant stirring at room temperature. This solution was then allowed to stir for 1 hour in a sealed container. 20% (W/W) solution of PAN in NMP was prepared separately and was added drop wise to the well-stirred PANI solution. The blend solution was allowed to stir with a mechanical stirrer for an additional 1 hour.

Various polymer blends with PANI content ranging from 10 wt% to 30 wt% were prepared by mixing different amount of 5% PANI solution and 20% PAN solution. Total concentration of the blend solutions were kept as 12.5%.

Polymeric nanofibers can be made using the electrospinning process, which has been described in the literature and patent [20-21]. Electrospinning uses a high electric field to draw a polymer solution from tip of a capillary toward a collector. A voltage is applied to the polymer solution, which causes a jet of the solution to be drawn toward a grounded collector. The fine jets dry to form polymeric fibers, which can be collected as a web.

Our electrospinning equipment used a variable high voltage power supply from Gamma High Voltage Research (USA). The applied voltage can be varied from 1- 30 kV. A 5-ml syringe was used and positive potential was applied to the polymer blend solution by attaching the electrode directly to the outside of the hypodermic needle with internal diameter of 0.3 mm. The collector screen was a 20×20 cm aluminum foil, which was placed 10 cm horizontally from the tip of the needle. The electrode of opposite polarity was attached to the collector. A metering syringe pump from New Era pump systems Inc. (USA) was used. It was responsible for supplying polymer solution with a constant rate of 20 μl/min.

Electrospinning was done in a temperature-controlled chamber and temperature of electrospinning environment was adjusted on 25, 50 and 75 °C. Schematic diagram of the electrospinning apparatus was shown in Figure 2. Factorial experiment was designed to investigate and identify the effects of parameters on fiber diameter and morphology. (Table 1)

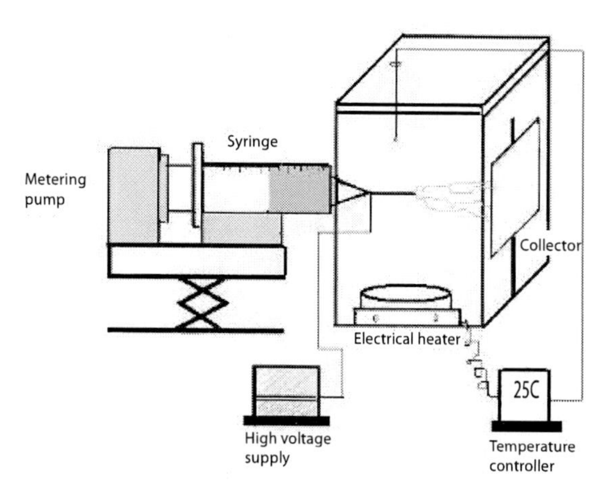

Figure 2. Schematic diagram of electrospinning apparatus.

Table 1. Factorial design of experiment

Factor	Factor level
PANI Content(wt%)	10,20,30
Electrospinning temperature(°C)	25,50,75
Applied voltage(kV)	20,25,30

Shear viscosities of the fluids were measured at shear rate of 500 sec-^{1}and 22 °C using a Brookfield viscometer (DVII+,USA).Fiber formation and morphology of the electrospun PANI/PAN fibers were determined using a scanning electron microscope (SEM) Philips XL-30A (Holland). Small section of the prepared samples was placed on SEM sample holder and then coated with gold by a BAL-TEC SCD 005 sputter coater. The diameter of electrospun fibers was measured with image analyzer software (manual microstructure distance measurement). For each experiment, average fiber diameter and distribution were determined from about 100 measurements of the random fibers. Electrical conductivity of the electrospun mats was measured by the standard four- probe method after doping with HCl vapor.

In this study, solution concentration (C), spinning distance (d), applied voltage (V), and volume flow rate (Q) were selected to be the most influential parameters. The next step is to choose the ranges over which these factors are varied. Process knowledge, which is a combination of practical experience and theoretical understanding, is required to fulfill this step. The aim is here to find an appropriate range for each parameter where dry, bead-free, stable, and continuous fibers without breaking up to droplets are obtained. This goal could be achieved by conducting a set of preliminary experiments while having the previous works in mind along with utilizing the reported relationships.

RESPONSE SURFACE METHODOLOGY

The mechanism of some scientific phenomena has been well understood and models depicting the physical behavior of the system have been drawn in the form of mathematical relationships. However, there are numerous processes at the moment which have not been sufficiently been understood to permit the theoretical approach. Response surface methodology (RSM) is a combination of mathematical and statistical techniques useful for empirical modeling and analysis of such systems. The application of RSM is in situations where several input variables are potentially influence some performance measure or quality characteristic of the process – often called responses. The relationship between the response (y) and k input variables ($\xi_1,\xi_2,...,\xi_k$) could be expressed in terms of mathematical notations as follows:

$$y = f(\xi_1, \xi_2, ..., \xi_k) \tag{3}$$

where the true response function f is unknown. It is often convenient to use coded variables ($x_1,x_2,..,x_k$) instead of natural (input) variables. The response function will then be:

$$y = f(x_1, x_2, ..., x_k) \tag{4}$$

Since the form of true response function f is unknown, it must be approximated. Therefore, the successful use of RSM is critically dependent upon the choice of appropriate function to approximate f. Low-order polynomials are widely used as approximating functions. First order (linear) models are unable to capture the interaction between parameters which is a form of curvature in the true response function. Second order (quadratic) models will be likely to perform well in these circumstances. In general, the quadratic model is in the form of:

$$y = \beta_0 + \sum_{j=1}^{k} \beta_j x_j + \sum_{j=1}^{k} \beta_{jj} x_j^2 + \sum_{i<j} \sum_{j=2}^{k} \beta_{ij} x_i x_j + \varepsilon \tag{5}$$

where ε is the error term in the model. The use of polynomials of higher order is also possible but infrequent. The βs are a set of unknown coefficients needed to be estimated. In order to do that, the first step is to make some observations on the system being studied. The model in Equation 5 may now be written in matrix notations as:

$$\mathbf{y} = \mathbf{X}\boldsymbol{\beta} + \boldsymbol{\varepsilon} \tag{6}$$

where $\mathbf{y}$ is the vector of observations, $\mathbf{X}$ is the matrix of levels of the variables, $\boldsymbol{\beta}$ is the vector of unknown coefficients, and $\boldsymbol{\varepsilon}$ is the vector of random errors. Afterwards, method of least squares, which minimizes the sum of squares of errors, is employed to find the estimators of the coefficients ($\hat{\boldsymbol{\beta}}$) through:

$$\hat{\boldsymbol{\beta}} = (\mathbf{X}'\mathbf{X})^{-1}\mathbf{X}'\mathbf{y} \tag{7}$$

The fitted model will then be written as:

$$\hat{\mathbf{y}} = \mathbf{X}\hat{\boldsymbol{\beta}} \tag{8}$$

Finally, response surfaces or contour plots are depicted to help visualize the relationship between the response and the variables and see the influence of the parameters [60], [61]. As you might notice, there is a close connection between RSM and linear regression analysis [62].

In this study RSM was employed to establish empirical relationships between four electrospinning parameters (solution concentration, spinning distance, applied voltage, and flow rate) and two responses (mean fiber diameter and standard deviation of fiber diameter). Coded variables were used to build the models. The choice of three levels for each factor in experimental design allowed us to take the advantage of quadratic models. Afterwards, the significance of terms in each model was investigated by testing hypotheses on individual coefficients and simpler yet more efficient models were obtained by eliminating statistically unimportant terms. Finally, the validity of the models was evaluated using the 15 test data. The analyses were carried out using statistical software Minitab 15.

RESULTS AND DISCUSSION

After the unknown coefficients (βs) were estimated by least squares method, the quadratic models for the mean fiber diameter (MFD) and standard deviation of fiber diameter (StdFD) in terms of coded variables are written as:

$$\begin{aligned} \text{MFD} = {} & 282.031 + 34.953x_1 + 5.622x_2 - 2.113x_3 + 9.013x_4 \\ & - 11.613x_1^2 - 4.304x_2^2 - 15.500x_3^2 \\ & - 0.414x_4^2 + 12.517x_1x_2 + 4.020x_1x_3 - 0.162x_1x_4 + 20.643x_2x_3 + 0.741x_2x_4 + 0.877x_3x_4 \end{aligned} \tag{9}$$

$$\begin{aligned} \text{StdFD} = {} & 36.1574 + 4.5788x_1 - 1.5536x_2 + 6.4012x_3 + 1.1531x_4 \\ & - 2.2937x_1^2 - 0.1115x_2^2 - 1.1891x_3^2 + 3.0980x_4^2 \\ & - 0.2088x_1x_2 + 1.0010x_1x_3 + 2.7978x_1x_4 + 0.1649x_2x_3 - 2.4876x_2x_4 + 1.5182x_3x_4 \end{aligned} \tag{10}$$

In the next step, a couple of very important hypothesis-testing procedures were carried out to measure the usefulness of the models presented here. First, the test for significance of the model was performed to determine whether there is a subset of variables which contributes significantly in representing the response variations. The appropriate hypotheses are:

$$\begin{aligned} & H_0 : \beta_1 = \beta_2 = \cdots = \beta_k \\ & H_1 : \beta_j \neq 0 \quad \text{for at least one } j \end{aligned} \tag{11}$$

The F statistics (the result of dividing the factor mean square by the error mean square) of this test along with the p-values (a measure of statistical significance, the smallest level of significance for which the null hypothesis is rejected) for both models are shown in Table 1.

Table 1. Summary of the results from statistical analysis of the models

	F	p-value	R^2	R^2_{adj}	R^2_{pred}
MFD	106.02	0.000	95.74%	94.84%	93.48%
StdFD	42.05	0.000	89.92%	87.78%	84.83%

The p-values of the models are very small (almost zero), therefore it could be concluded that the null hypothesis is rejected in both cases suggesting that there are some significant terms in each model. There are also included in Table 1, the values of R^2, R^2_{adj}, and R^2_{pred}. R^2 is a measure for the amount of response variation which is explained by variables and will always increase when a new term is added to the model regardless of whether the inclusion of the additional term is statistically significant or not. R^2_{adj} is the adjusted form of R^2 for the number of terms in the model; therefore it will increase only if the new terms improve the model and decreases if unnecessary terms are added. R^2_{pred} implies how well the model predicts the response for new observations, whereas R^2 and R^2_{adj} indicate how well the model fits the experimental data. The R^2 values demonstrate that 95.74% of MFD and 89.92% of StdFD are explained by the variables. The R^2_{adj} values are 94.84% and 87.78% for MFD and StdFD respectively, which account for the number of terms in the models. Both R^2 and R^2_{adj} values indicate that the models fit the data very well. The slight difference between the values of R^2 and R^2_{adj} suggests that there might be some insignificant terms in the models. Since the R^2_{pred} values are so close to the values of R^2 and R^2_{adj}, models does not appear to be overfit and have very good predictive ability.

The second testing hypothesis is evaluation of individual coefficients, which would be useful for determination of variables in the models. The hypotheses for testing of the significance of any individual coefficient are:

$$H_0 : \beta_j = 0$$
$$H_1 : \beta_j \neq 0 \tag{12}$$

Table 2. The test on individual coefficients for the model of mean fiber diameter (MFD)

Term (coded)	Coef	T	p-value
Constant	282.031	102.565	0.000
C	34.953	31.136	0.000
d	5.622	5.008	0.000
V	-2.113	-1.882	0.064
Q	9.013	8.028	0.000
C^2	-11.613	-5.973	0.000
d^2	-4.304	-2.214	0.030
V^2	-15.500	-7.972	0.000
Q^2	-0.414	-0.213	0.832
Cd	12.517	9.104	0.000
CV	4.020	2.924	0.005
CQ	-0.162	-0.118	0.906
dV	20.643	15.015	0.000
dQ	0.741	0.539	0.592
VQ	0.877	0.638	0.526

Table 3. The test on individual coefficients for the model of standard deviation of fiber diameter (StdFD)

Term (coded)	Coef	T	*p*-value
Constant	36.1574	39.381	0.000
C	4.5788	12.216	0.000
D	-1.5536	-4.145	0.000
V	6.4012	17.078	0.000
Q	1.1531	3.076	0.003
C^2	-2.2937	-3.533	0.001
d^2	-0.1115	-0.172	0.864
V^2	-1.1891	-1.832	0.072
Q^2	3.0980	4.772	0.000
Cd	-0.2088	-0.455	0.651
CV	1.0010	2.180	0.033
CQ	2.7978	6.095	0.000
dV	0.1649	0.359	0.721
dQ	-2.4876	-5.419	0.000
VQ	1.5182	3.307	0.002

The model might be more efficient with inclusion or perhaps exclusion of one or more variables. Therefore the value of each term in the model is evaluated using this test, and then eliminating the statistically insignificant terms, more efficient models could be obtained. The results of this test for the models of MFD and StdFD are summarized in Table 2 and Table 3 respectively. *T* statistic in these tables is a measure of the difference between an observed statistic and its hypothesized population value in units of standard error.

As depicted, the terms related to Q^2, CQ, dQ, and VQ in the model of MFD and related to d^2, Cd, and dV in the model of StdFD have very high *p*-values, therefore they do not contribute significantly in representing the variation of the corresponding response. Eliminating these terms will enhance the efficiency of the models. The new models are then given by recalculating the unknown coefficients in terms of coded variables in equations (13) and (14), and in terms of natural (uncoded) variables in equations (15), (16).

$$\begin{aligned} MFD = {} & 281.755 + 34.953x_1 + 5.622x_2 - 2.113x_3 + 9.013x_4 \\ & - 11.613x_1^2 - 4.304x_2^2 - 15.500x_3^2 \\ & + 12.517x_1x_2 + 4.020x_1x_3 + 20.643x_2x_3 \end{aligned} \tag{13}$$

$$\begin{aligned} StdFD = {} & 36.083 + 4.579x_1 - 1.554x_2 + 6.401x_3 + 1.153x_4 \\ & - 2.294x_1^2 - 1.189x_3^2 + 3.098x_4^2 \\ & + 1.001x_1x_3 + 2.798x_1x_4 - 2.488x_2x_4 + 1.518x_3x_4 \end{aligned} \tag{14}$$

$$\begin{aligned} MFD = {} & 10.3345 + 48.7288\,C - 22.7420\,d + 7.9713\,V + 90.1250\,Q \\ & - 2.9033\,C^2 - 0.1722\,d^2 - 0.6120\,V^2 \\ & + 1.2517\,Cd + 0.4020\,CV + 0.8257\,dV \end{aligned} \tag{15}$$

$$\begin{aligned} StdFD = {} & -1.8823 + 7.5590\,C + 1.1818\,d + 1.2709\,V - 300.3410\,Q \\ & - 0.5734\,C^2 - 0.0476\,V^2 + 309.7999\,Q^2 \\ & + 0.1001\,CV + 13.9892\,CQ - 4.9752\,dQ + 3.0364\,VQ \end{aligned} \tag{16}$$

The results of the test for significance as well as R^2, R^2_{adj}, and R^2_{pred} for the new models are given in Table 4. It is obvious that the *p*-values for the new models are close to zero indicating the existence of some significant terms in each model. Comparing the results of this table with Table 1, the *F* statistic increased for the new models, indicating the improvement of the models after eliminating the insignificant terms. Despite the slight decrease in R^2, the values of R^2_{adj}, and R^2_{pred} increased substantially for the new models. As it was mentioned earlier in the paper, R^2 will always increase with the number of terms in the model. Therefore, the smaller R^2 values were expected for the new models, due to the fewer terms. However, this does not necessarily suggest that the pervious models were more efficient. Looking at the tables, R^2_{adj}, which provides a more useful tool for comparing the explanatory power of models with different number of terms, increased after eliminating the unnecessary variables. Hence, the new models have the ability to better explain the experimental data. Due to higher R^2_{pred}, the new models also have higher prediction ability. In other words, eliminating the insignificant terms results in simpler models which not only present the experimental data in superior form, but also are more powerful in predicting new conditions. In the study conducted by Yördem et al. [36], despite high reported R^2 values, the presented models seem to be inefficient and uncertain. Some terms in the models had very high *p*-values. For instance, in modeling the mean fiber diameter, *p*-value as high as 0.975 was calculated for cubic concentration term at spinning distance of 16 *cm*, where half of the terms had *p*-values more than 0.8. This results in low R^2_{pred} values which were not reported in their study and after calculating by us, they were found to be almost zero in many cases suggesting the poor prediction ability of their models.

Table 4. Summary of the results from statistical analysis of the models after eliminating the insignificant terms

	F	*p*-value	R^2	R^2_{adj}	R^2_{pred}
MFD	155.56	0.000	95.69%	95.08%	94.18%
StdFD	55.61	0.000	89.86%	88.25%	86.02%

Table 5. The test on individual coefficients for the model of mean fiber diameter (MFD) after eliminating the insignificant terms

Term (coded)	Coef	T	*p*-value
Constant	281.755	118.973	0.000
C	34.953	31.884	0.000
d	5.622	5.128	0.000
V	-2.113	-1.927	0.058
Q	9.013	8.221	0.000
C^2	-11.613	-6.116	0.000
d^2	-4.304	-2.267	0.026
V^2	-15.500	-8.163	0.000
Cd	12.517	9.323	0.000
CV	4.020	2.994	0.004
dV	20.643	15.375	0.000

The test for individual coefficients was performed again for the new models. The results of this test are summarized in Table 5 and Table 6. This time, as it was anticipated, no terms had higher *p*-value than expected, which need to be eliminated. Here is another advantage of removing unimportant terms. The values of *T* statistic increased for the terms already in the models implying that their effects on the response became stronger.

After developing the relationship between parameters, the test data were used to investigate the prediction ability of the models. Root mean square errors (RMSE) between the calculated responses (C_i) and real responses (R_i) were determined using equation (17) for experimental data as well as test data for the sake of evaluation of both MFD and StdFD models and the results are listed in Table 7. The models present acceptable RMSE values for test data indicating the ability of the models to generalize well the experimental data to predicting new conditions. Although the values of RMSE for the test data are slightly higher than experimental data, these small discrepancies were expected since it is almost impossible for an empirical model to express the test data as well as experimental data and higher errors are often obtained when new data are presented to the models. Hence, the results imply the acceptable prediction ability of the models.

Table 6. The test on individual coefficients for the model of standard deviation of fiber diameter (StdFD) after eliminating the insignificant terms

Term (coded)	Coef	T	*p*-value
Constant	36.083	45.438	0.000
C	4.579	12.456	0.000
d	-1.554	-4.226	0.000
V	6.401	17.413	0.000
Q	1.153	3.137	0.003
C^2	-2.294	-3.602	0.001
V^2	-1.189	-1.868	0.066
Q^2	3.098	4.866	0.000
CV	1.001	2.223	0.029
CQ	2.798	6.214	0.000
dQ	-2.488	-5.525	0.000
VQ	1.518	3.372	0.001

$$\mathrm{RMSE} = \sqrt{\frac{\sum_{i=1}^{n}(C_i - R_i)^2}{n}} \tag{17}$$

Table 7. RMSE values of the models for the experimental and test data

	Experimental data	Test data
MFD	7.489	10.647
StdFD	2.493	2.890

Response Surfaces for Mean Fiber Diameter

Solution Concentration

A monotonic increase in MFD with concentration was observed in this study as shown in Figure (a), (b), and (c) which concurs with the previous observations [25], [31], [63]- [65]. The concentration effect was more pronounced at further spinning distances (Figure (a)). This could be attributed to the twofold effect of distance. At low concentrations, higher solvent content in the solution and longer distance provides more time not only to stretch the jet in the electric field but also to evaporate the solvent, thereby encouraging thinner fiber formation. At higher concentrations, however, there are extensive polymer chain entanglements resulting in higher viscoelastic forces which tend to resist against the electrostatic stretching force. On the other hand, increasing the spinning distance will reduce the electric field strength ($E=V/d$) causing the electrostatic force to decrease. As a result, increasing MFD with concentration gains more momentum at longer spinning distances. Higher applied voltages also accelerate the concentration impact on MFD (Figure (b)) which may be ascribed to the two fold effect of voltage. At higher voltages, where the electric field is strong and dominant factor, increasing polymer concentration tends to encourage the effect of voltage on mass flow rate of polymer. Hence, more solution could be removed from the tip of the needle resulting in further increase in MFD. No combined effect between solution concentration and volume flow rate was observed as depicted in Figure (c). Therefore, concentration had interactions with spinning distance and applied voltage which had been suggested by the existence of terms *Cd* and *CV* in the model of MFD. Recall that the term *CQ* was statistically insignificant and therefore had been removed from the model of MFD.

Spinning Distance

The impact of spinning distance on MFD is illustrated in Figure (a), (d), and (e). As it is depicted in these figures, the effect of spinning distance is not always the same. Spinning distance has a twofold effect on electrospun fiber diameter. Varying the distance has a direct influence on the jet flight time as well as electric field strength. Longer spinning distance will provide more time for the jet to stretch in the electric field before it is deposited on the collector. Furthermore, solvents will have more time to evaporate. Hence, the fiber diameter will be prone to decrease. On the other hand, increasing the spinning distance, the electric field strength will decrease ($E=V/d$) resulting in less acceleration hence stretching of the jet which leads to thicker fiber formation. The balance between these two effects will determine the final fiber diameter. Increase in fiber diameter [64], [67], [68] as well as decrease in fiber diameter [31] with increasing spinning distance was reported in the literature. There were also some cases in which spinning distance did not have a significant influence on fiber diameter [63], [69]-[71]. As mentioned before, there will be more chain entanglements at higher concentrations resulting in an increase in viscoelastic force. Furthermore, the longer the distance, the lower is the electric field strength. Hence, the electrostatic stretching force, which has now become weaker, will be dominated easier by the viscoelastic force. As a result, the effect of spinning distance on fiber diameter is more highlighted, rendering higher

MFD (Figure (a)). The effect of spinning distance will also alter at different applied voltages (Figure (d)). At low voltages, longer spinning distance brought about thinner fiber formation, whereas at high voltages, the effect of spinning distance was totally reversed and fibers with thicker diameters were obtained at longer distances. It is supposed that at low voltages, the stretching time becomes the dominant factor. Hence, longer spinning distance, which gives more time for jet stretching and thinning and solvent evaporation, will result in fibers with smaller diameters. At high voltages, however, the electric field strength is high and dominant. Therefore, increasing the distance, which reduces the electric field, causes an increase in fiber diameter. The function of spinning distance was observed to be independent from volume flow rate for MFD (Figure (e)). The interaction of spinning distance with solution concentration and applied voltage demonstrated in Figure (a) and (d), proved the existence of terms *Cd* and *dV* in the model of MFD.

Applied Voltage

Figure 3 (b), (d), and (f) show the effect of applied voltage on MFD. Increasing the voltage resulted in an increase followed by a decrease in MFD. Applied voltage has two major different effects on fiber diameter. Firstly, increasing the applied voltage will increase the electric field strength and larger electrostatic stretching force causes the jet to accelerate more in the electric field, thereby favoring thinner fiber formation. Secondly, since charge transport is only carried out by the flow of polymer in the electrospinning process [72], increasing the voltage would induce more surface charges on the jet. Subsequently, the mass flow rate from the needle tip to the collector will increase, say the solution will be drawn more quickly from the tip of the needle causing fiber diameter to increase. Combination of these two effects will determine the final fiber diameter. Hence, increasing applied voltage may decrease [73]-[75], increase [63], [64], [68] or may not change [25], [31], [69], [76] the fiber diameter. According to the given explanation, at low voltages, where the electric field strength is low, the effect of mass of solution could be dominant. Therefore, fiber diameter increases when the applied voltage rises. However, as the voltage exceeds a limit, the electric field will be high enough to be a determining factor. Hence, fiber diameter decreases as the voltage increases. The effect of voltage on MFD was influenced by solution concentration to some extent (Figure 3 (b)). At high concentrations, the increase in fiber diameter with voltage was more pronounced. This could be attributed to the fact that the effect of mass of solution will be more important for the solutions of higher concentrations. The change in fiber diameter as a function of voltage is dramatically influenced by spinning distance (Figure 3 (d)). At a short distance, the electric field is high and dominant factor. Therefore, increasing applied voltage, which strengthens the electric field, results in a decrease in fiber diameter. Whereas, at long distances where the electric field is low, the effect of mass of solution would be determining factor according to which fiber diameter increased with applied voltage. The effect of applied voltage on MFD is found to be independent from volume flow rate (Figure 3 (f)). It is quite apparent that there is a huge interaction between applied voltage and spinning distance, a slight interaction between applied voltage and solution concentration and no interaction between applied voltage and volume flow rate which is in agreement with the presence of *CV* and *dV* and absence of *VQ* in the model of MFD.

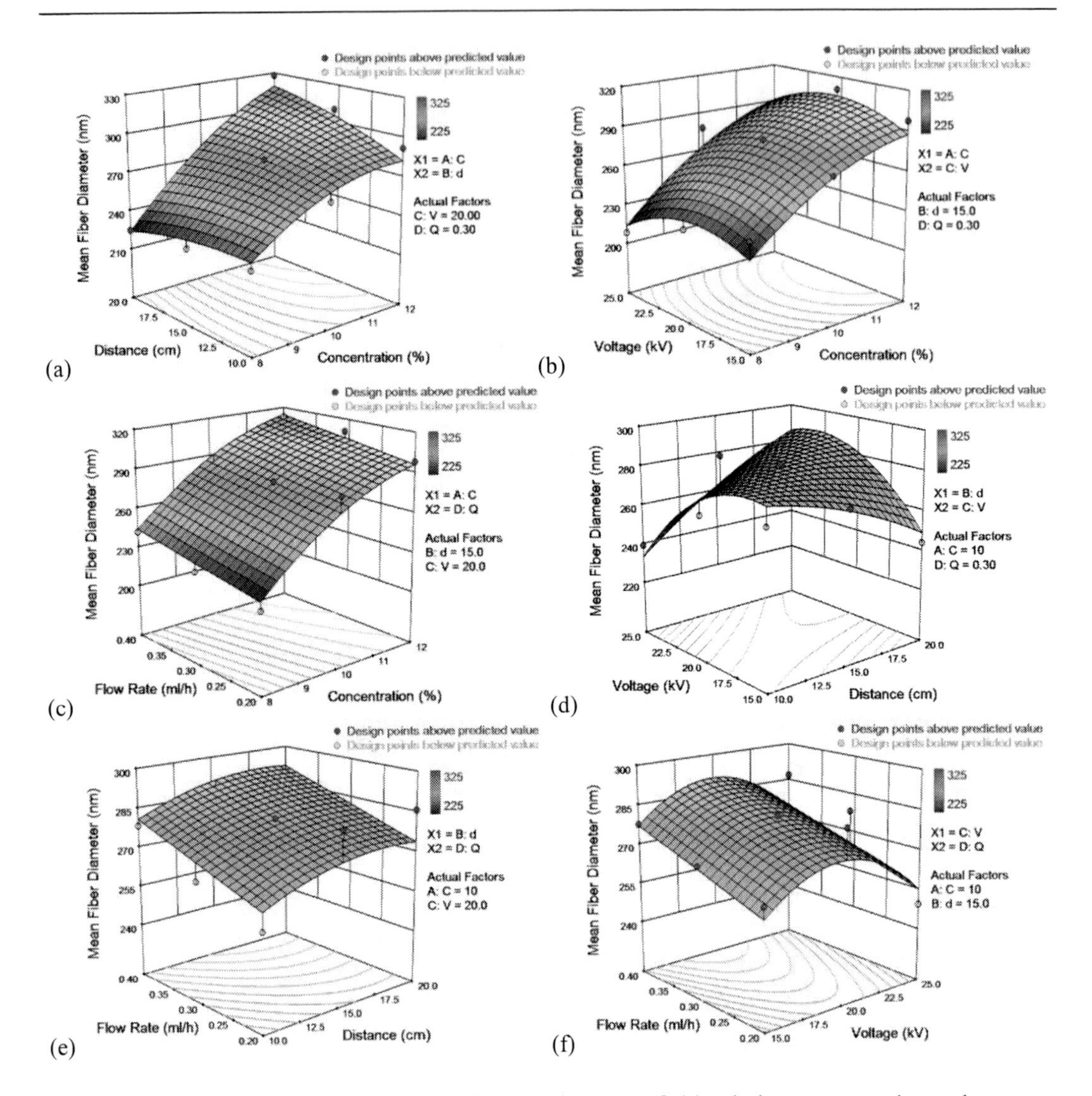

Figure 3. Response surfaces for mean fiber diameter in terms of: (a) solution concentration and spinning distance, (b) solution concentration and applied voltage, (c) solution concentration and flow rate, (d) spinning distance and applied voltage, (e) spinning distance and flow rate, (f) applied voltage and flow rate

Volume Flow Rate

It was suggested that a minimum value for solution flow rate is required to form the drop of polymer at the tip of the needle for the sake of maintaining a stable Taylor cone [77]. Hence, flow rate could affect the morphology of electrospun nanofibers such as fiber diameter. Increasing the flow rate, more amount of solution is delivered to the tip of the needle enabling the jet to carry the solution away faster. This could bring about an increase in the jet diameter favoring thicker fiber formation. In this study, the MFD slightly increased with volume flow rate (Figure 2 (c), (e), and (f)) which agrees with the previous researches [31], [77]-[79]. Flow rate was also found to influence MFD independent from solution concentration, applied voltage, and spinning distance as suggested earlier by the absence of *CQ*, *dQ*, and *VQ* in the model of MFD.

RESPONSE SURFACES FOR STANDARD DEVIATION OF FIBER DIAMETER

Solution Concentration

As depicted in Figure 3 (a), (b), and (c), StdFD increased with concentration which is in agreement with the previous observations [25], [34], [63], [66], [31], [68], [80], [81]. Increasing the polymer concentration, the macromolecular chain entanglements increase, prompting a greater difficulty for the jet to stretch and split. This could result in less uniform fibers (higher StdFD). Concentration affected StdFD regardless of spinning distance (Figure 4 (a)), suggesting that there was no interaction between these two parameters (absence of *Cd* in the model of StdFd). At low applied voltages, the formation of more uniform fibers with decreasing the concentration was facilitated. In agreement with existence of the term *CV* in the model of StdFd, solution concentration was found to have a slight interaction with applied voltage (Figure 3 (b)). The curvature of the surface in Figure 3 (c) suggested that there was a noticeable interaction between concentration and flow rate and this agrees with the presence of the term *CQ* in the model of StdFD.

Spinning Distance

More uniform fibers (lower StdFD) were obtained with increasing the spinning distance as shown in Figure (a), (d), and (e). At longer spinning distance, more time is provided for jet flying from the tip of the needle to the collector and solvent evaporation. Therefore, jet stretching and solvent evaporation is carried out more gently resulting in more uniform fibers. Our finding is consistent with the trend observed by Zhao et al. [81]. Spinning distance influenced StdFD regardless of solution concentration and applied voltage (Figure (a) and (d)) indicating that no interaction exists between these variables as could be inferred from the model of StdFD. However, the interaction of spinning distance with volume flow rate is obvious (Figure (e)). The presence of *dQ* in the model of StdFD proves this observation. The effect of spinning distance is more highlighted at higher flow rates. This could be attributed to the fact that more amount of solution is delivered to the tip of the needle at higher flow rates; therefore the threads will require more time to dry. If the distance is high enough to provide the sufficient time, uniform fibers will be formed. Decreasing the distance, there will be less time for solvent to evaporate favoring the production of non-uniform fibers (high StdFD).

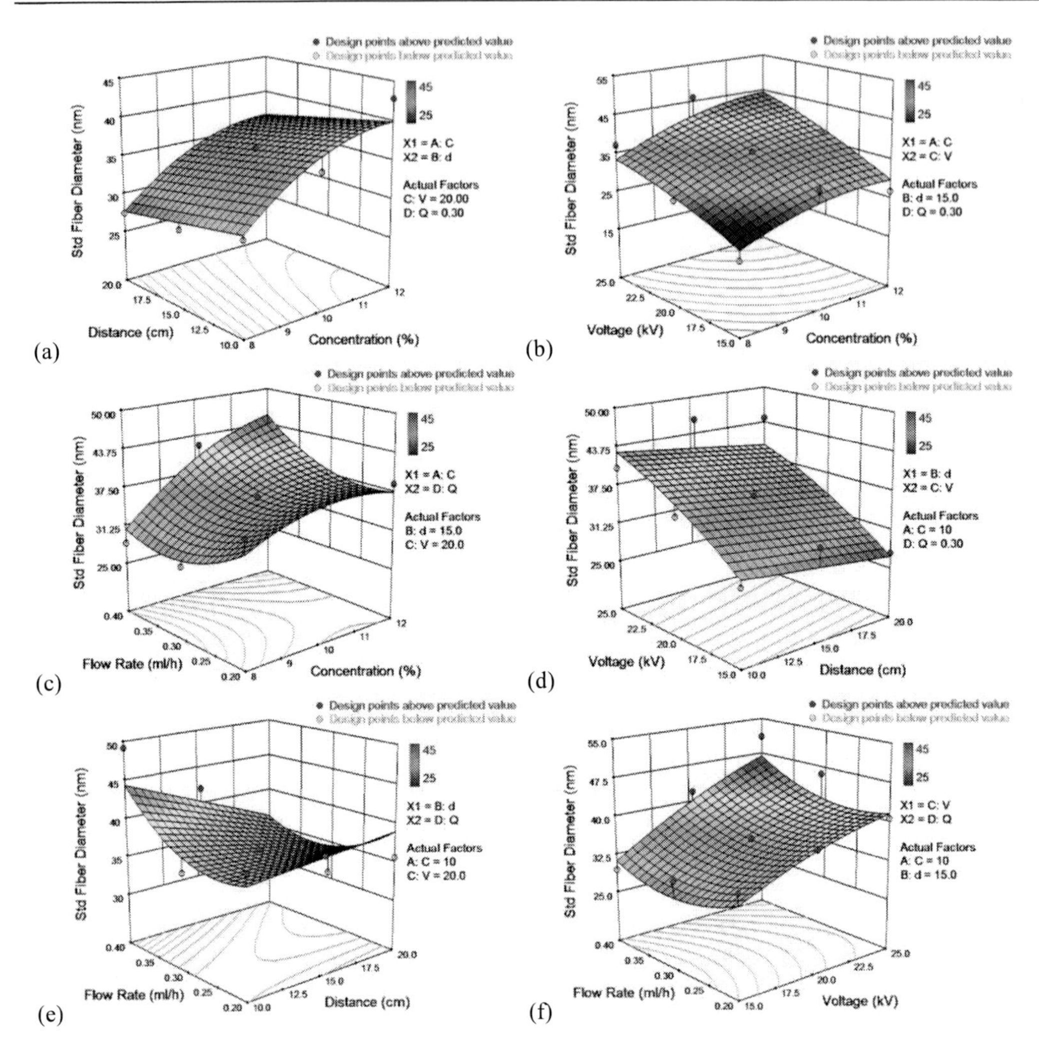

Figure 4. Response surfaces for standard deviation of fiber diameter in terms of: (a) solution concentration and spinning distance, (b) solution concentration and applied voltage, (c) solution concentration and flow rate, (d) spinning distance and applied voltage, (e) spinning distance and flow rate, (f) applied voltage and flow rate

Applied Voltage

StdFD was found to increase with applied voltage (Figure 3 (b), (d), and (f)) as observed in other studies [63], [64], [68], [81]. Increasing the applied voltage causes the effect of the electric field on the charged jet to increase. Hence, the flight speed of the jet increases, shortening the time that the jet travels towards the collector. As a result, less time is provided for jet stretching and thinning and also solvent evaporation. This may result in formation of less uniform fibers (higher StdFD). The effect of applied voltage on StdFD is influenced by solution concentration as shown in Figure (b), implying the interaction of voltage with concentration which was earlier addressed in the paper by the presence of the corresponding term in the model of StdFD. At low concentrations, the formation of uniform fibers with decreasing the applied voltage was facilitated. No interaction was observed between applied voltage and spinning distance (Figure (d)) as suggested by the absence of the term dV in the model of StdFD. Figure (f) shows a slight interaction of voltage with flow rate, which concurs with the existence of VQ in the model of StdFD.

Volume Flow Rate

As demonstrated in Figure (c), (e), and (f), the uniformity of fibers increased (StdFD decreased), reached to an optimum value and then decreased (StdFD increased) by increasing of the flow rate. When the flow rate is low, the amount of solution fed to the tip of the needle is not sufficient, whereas an excess amount of solution is delivered to the tip of the needle at high flow rates. Therefore, unstable jets are formed in the two extremes resulting in the production of non-uniform fibers. The impact of flow rate on StdFD is influenced by solution concentration, applied voltage, and spinning distance. This observation indicates the interaction between flow rate and other variables as demonstrated by the terms CQ, dQ, and VQ in the model of StdFD. Increasing the solution concentration favored the formation of non-uniform fibers at high flow rates (Figure (c)) which is probably the outcome of greater difficulty of solution removal. The effect of flow rate on StdFD was more pronounced as the spinning distance decreased (Figure (e)). The shorter the distance, the fewer the time provided to the jet to thin and dry. Therefore, at high flow rates, more amount of solution is delivered with insufficient flying time, results in formation of less uniform fibers. High applied voltage encouraged the increase in StdFD at fast flow rates as depicted in Figure (f).

Conclusion

The simultaneous effects of four processing variables including solution concentration, applied voltage, spinning distance, and volume flow rate on MFD and StdFD were investigated quantitatively and qualitatively. The appropriate range of parameters where dry, bead-free, and continuous fibers without breaking up to droplets are formed, were selected by referring to the literature along with conducting a series of preliminary experiments. A full factorial experimental design at three levels of each factor (3^4 design) was carried out. Moreover, 15 treatments inside the design space were selected as test set for evaluating the prediction ability of the models. PVA Nanofibers were then prepared for experimental and test sets through the electrospinning method. After that, MFD and StdFD were determined from SEM micrograph of each sample. RSM was used to establish quadratic models for MFD and StdFD. The test for significance of the coefficients demonstrated that the terms Q^2, CQ, dQ, and VQ in the model of MFD and d^2, Cd, and dV in the model of StdFD were not of important value in representing the responses. Eliminating these terms, simpler yet more efficient models were obtained which not only explained the experimental data in a better manner, but also had more prediction ability. Afterwards, in order to show the generalization ability of the models for predicting new conditions, the test set was used. Low RMSE of test set for MFD and StdFD were obtained indicating the good prediction ability of the models.

Finally, in order to qualitatively study the effects of variables on MFD and StdFD, response surface plots were generated using the obtained relationships. For MFD:

1. Increasing solution concentration, MFD increased rigorously. The effect of concentration was more pronounced at longer spinning distance and also at higher applied voltage.
2. The effect of spinning distance on MFD changed depending on solution concentration and applied voltage. At low applied voltages, MFD decreased as the spinning distance became longer, whereas higher MFD resulted with lengthening the spinning distance when the applied voltage was high. Increasing the solution concentration tended to assist the formation of thicker fibers at longer spinning distance.
3. Rising the applied voltage, MFD was observed to first increase and then decrease. High solution concentrations partly and long spinning distances largely favored the increase of MFD with applied voltage.
4. MFD slightly increased with flow rate. The impact of flow rate on MFD was unrelated to the other variables.

For StdFD:

1. The higher the solution concentration, the less uniform fibers (higher StdFD) was formed. Low applied voltages facilitated the formation of more uniform fibers (lower StdFD) with decreasing the concentration. The increase of StdFD with concentration gained momentum at high flow rates.
2. Longer spinning distance resulted in more uniform fibers (lower StdFD). The effect of spinning distance was more pronounced at higher flow rates.
3. Rising the applied voltage increased StdFD. Low concentrations facilitated the formation of uniform fibers (high StdFD) with decreasing the applied voltage.
4. Flow rate was found to have a significant impact on uniformity of fibers (StdFD). As flow rate increased, StdFD decreased and then increased. Higher solution concentration, higher applied voltage, and shorter spinning distance encouraged the formation of non-uniform fibers (high StdFD) at fast flow rates.

REFERENCES

[1] G.I. Taylor, *Proc. Roy. Soc. London.*, 313, 453 (1969).

[2] J. Doshi and D.H. Reneker, *J. Electrostatics*, 35, 151 (1995).

[3] H. Fong and D.H. Reneker, *Electrospinning and the formation of nanofibers*, in: D.R. Salem (Ed.), *Structure formation in polymeric fibers*, Hanser, Cincinnati (2001).

[4] D. Li and Y. Xia, *Adv. Mater.*, 16, 1151 (2004).

[5] R. Derch, A. Greiner and J.H. Wendorff, *Polymer nanofibers prepared by electrospinning*, in: J.A. Schwarz, C.I. Contescu and K. Putyera (Eds.), *Dekker encyclopedia of nanoscience and nanotechnology*, CRC, New York (2004).

[6] A.K. Haghi and M. Akbari, *Phys. Stat. Sol. A*, 204, 1830 (2007).

[7] P.W. Gibson, H.L. Schreuder-Gibson and D. Rivin, *AIChE J.*, 45, 190 (1999).
[8] M. Ziabari, V. Mottaghitalab and A.K. Haghi, *Korean J. Chem. Eng.*, 25, 923 (2008).
[9] Z.M. Huang, Y.Z. Zhang, M. Kotaki and S. Ramakrishna, *Compos. Sci. Technol.*, 63, 2223 (2003).
[10] M. Li, M.J. Mondrinos, M.R. Gandhi, F.K. Ko, A.S. Weiss and P.I. Lelkes, *Biomaterials*, 26, 5999 (2005).
[11] E.D. Boland, B.D. Coleman, C.P. Barnes, D.G. Simpson, G.E. Wnek and G.L. Bowlin, *Acta. Biomater.*, 1, 115 (2005).
[12] J. Lannutti, D. Reneker, T. Ma, D. Tomasko and D. Farson, *Mater. Sci. Eng. C*, 27, 504 (2007).
[13] J. Zeng, L. Yang, Q. Liang, X. Zhang, H. Guan, C. Xu, X. Chen and X. Jing, *J. Control. Release*, 105, 43 (2005).
[14] E.R. Kenawy, G.L. Bowlin, K. Mansfield, J. Layman, D.G. Simpson, E.H. Sanders and G.E. Wnek, *J. Control. Release*, 81, 57 (2002).
[15] M.S. Khil, D.I. Cha, H.-Y. Kim, I.-S. Kim and N. Bhattarai, *J. Biomed. Mater. Res. Part B: Appl. Biomater.*, 67, 675 (2003).
[16] B.M. Min, G. Lee, S.H. Kim, Y.S. Nam, T.S. Lee and W.H. Park, *Biomaterials*, 25, 1289 (2004).
[17] X.H. Qin and S.Y. Wang, *J. Appl. Polym. Sci.*, 102, 1285 (2006).
[18] H.S. Park and Y.O. Park, *Korean J. Chem. Eng.*, 22, 165 (2005).
[19] J.S. Kim and D.H. Reneker, *Poly. Eng. Sci.*, 39, 849 (1999).
[20] S.W. Lee, S.W. Choi, S.M. Jo, B.D. Chin, D.Y. Kim and K.Y. Lee, *J. Power Sources*, 163, 41 (2006).
[21] C. Kim, *J. Power Sources*, 142, 382 (2005).
[22] N.J. Pinto, A.T. Johnson, A.G. MacDiarmid, C.H. Mueller, N. Theofylaktos, D.C. Robinson and F.A. Miranda, *Appl. Phys. Lett.*, 83, 4244 (2003).
[23] D. Aussawasathien, J.-H. Dong and L. Dai, *Synthetic Met.*, 54, 37 (2005).
[24] S.-Y. Jang, V. Seshadri, M.-S. Khil, A. Kumar, M. Marquez, P.T. Mather and G.A. Sotzing, *Adv. Mater.*, 17, 2177 (2005).
[25] S.-H. Tan, R. Inai, M. Kotaki and R. Ramakrishna, *Polymer*, 46, 6128 (2005).
[26] A, Ziabicki, *Fundamentals of fiber formation: The science of fiber spinning and drawing*, Wiley, New York (1976).
[27] A. Podgóski, A. Bałazy and L. Gradoń, *Chem. Eng. Sci.*, 61, 6804 (2006).
[28] B. Ding, M. Yamazaki and S. Shiratori, *Sens. Actuators B*, 106, 477 (2005).
[29] J.R. Kim, S.W. Choi, S.M. Jo, W.S. Lee and B.C. Kim, *Electrochim. Acta*, 50, 69 (2004).
[30] L. Moroni, R. Licht, J. de Boer, J.R. de Wijn and C.A. van Blitterswijk, *Biomaterials*, 27, 4911 (2006).
[31] T. Wang and S. Kumar, *J. Appl. Polym. Sci.*, 102, 1023 (2006).
[32] W. Cui, X. Li, S. Zhou and J. Weng, *J. Appl. Polym. Sci.*, 103, 3105 (2007).
[33] S. Sukigara, M. Gandhi, J. Ayutsede, M. Micklus and F. Ko, *Polymer*, 45, 3701 (2004).
[34] S.Y. Gu, J. Ren and G.J. Vancso, *Eur. Polym. J.*, 41, 2559 (2005).
[35] S.Y. Gu and J. Ren, *Macromol. Mater. Eng.*, 290, 1097 (2005).
[36] O.S. Yördem, M. Papila and Y.Z. Menceloğlu, *Mater. Design*, 29, 34 (2008).
[37] I. Sakurada, *Polyvinyl Alcohol Fibers*, CRC, New York (1985).

[38] F.L. Marten, *Vinyl alcohol polymers*, in: H.F. Mark (Ed.), *Encyclopedia of polymer science and technology*, 3rd ed., vol. 8, Wiley (2004).
[39] Y.D. Kwon, S. Kavesh and D.C. Prevorsek, US Patent, 4,440,711 (1984).
[40] S. Kavesh and D.C. Prevorsek, US Patent, 4,551,296 (1985).
[41] H. Tanaka, M. Suzuki and F. Uedo, US Patent, 4,603,083 (1986).
[42] G. Paradossi, F. Cavalieri, E. Chiessim, C. Spagnoli and M.K. Cowman, *J. Mater. Sci.: Mater. Med.*, 14, 687 (2003).
[43] G. Zheng-Qiu, X. Jiu-Mei and Z. Xiang-Hong, *Biomed. Mater. Eng.*, 8, 75 (1998).
[44] M. Oka, K. Ushio, P. Kumar, K. Ikeuchi, S.H. Hyon, T. Nakamura and H. Fujita, *P. I. Mech. Eng. H*, 214, 59 (2000).
[45] K. Burczak, E. Gamian and A. Kochman, *Biomaterials*, 17, 2351 (1996).
[46] J.K. Li, N. Wang and X.S. Wu, *J. Control. Release*, 56, 117 (1998).
[47] A.S. Hoffman, *Adv. Drug Delivery Rev.*, 43, 3 (2002).
[48] J. Zeng, A. Aigner, F. Czubayko, T. Kissel, J.H. Wendorff and A. Greiner, *Biomacromolecules*, 6, 1484 (2005).
[49] K.H. Hong, *Polym. Eng. Sci.*, 47, 43 (2007).
[50] M. Ziabari, V. Mottaghitalab and A.K. Haghi, *Korean J. Chem. Eng.*, 25, 919 (2008).
[51] M. Ziabari, V. Mottaghitalab and A.K. Haghi, *Braz. J. Chem. Eng.*, 26, 53 (2009).
[52] M. Ziabari, V. Mottaghitalab, S.T. McGovern and A.K. Haghi, *Nanoscale Res. Lett.*, 2, 597 (2007).
[53] M. Ziabari, V. Mottaghitalab and A.K. Haghi, *Korean J. Chem. Eng.*, 25, 905 (2008).
[54] L.H. Sperling, *Introduction to physical polymer science*, 4th ed., Wiley, New Jersey (2006).
[55] J.C.J.F. Tacx, H.M. Schoffeleers, A.G.M. Brands and L. Teuwen, *Polymer*, 41, 947 (2000).
[56] F.K. Ko, *Nanofiber technology*, in: Y. Gogotsi (Ed.), *Nanomaterials handbook*, CRC, Boca Raton (2006).
[57] A. Koski, K. Yim and S. Shivkumar, *Mater. Lett.*, 58, 493 (2004).
[58] D.C. Montgomery, *Design and analysis of experiments*, 5th ed., Wiley, New York (1997).
[59] A. Dean and D. Voss, *Design and analysis of experiments*, Springer, New York (1999).
[60] G.E.P. Box and N.R. Draper, *Response surfaces, mixtures, and ridge analyses*, Wiley, New Jersey (2007).
[61] K.M. Carley, N.Y. Kamneva and J. Reminga, Response surface methodology, *CASOS Technical Report*, CMU-ISRI-04-136 (2004).
[62] S. Weisberg, *Applied linear regression*, 3rd ed., Wiley, New Jersey (2005).
[63] C. Zhang, X. Yuan, L. Wu, Y. Han and J. Sheng, *Eur. Polym. J.*, 41, 423 (2005).
[64] Q. Li, Z. Jia, Y. Yang, L. Wang and Z. Guan, Preparation and properties of poly(vinyl alcohol) nanofibers by electrospinning, *Proceedings of IEEE International Conference on Solid Dielectrics*, Winchester, U.K. (2007).
[65] C. Mit-uppatham, M. Nithitanakul and P. Supaphol, *Macromol. Chem. Phys.*, 205, 2327 (2004).
[66] Y.J. Ryu, H.Y. Kim, K.H. Lee, H.C. Park and D.R. Lee, *Eur. Polym. J.*, 39, 1883 (2003).

[67] T. Jarusuwannapoom, W. Hongrojjanawiwat, S. Jitjaicham, L. Wannatong, M. Nithitanakul, C. Pattamaprom, P. Koombhongse, R. Rangkupan and P. Supaphol, *Eur. Polym. J.*, 41, 409 (2005).
[68] S.C. Baker, N. Atkin, P.A. Gunning, N. Granville, K. Wilson, D. Wilson and J. Southgate, *Biomaterials*, 27, 3136 (2006).
[69] S. Sukigara, M. Gandhi, J. Ayutsede, M. Micklus and F. Ko, *Polymer*, 44, 5721 (2003).
[70] X. Yuan, Y. Zhang, C. Dong and J. Sheng, *Polym. Int.*, 53, 1704 (2004).
[71] C.S. Ki, D.H. Baek, K.D. Gang, K.H. Lee, I.C. Um and Y.H. Park, *Polymer*, 46, 5094 (2005).
[72]] J.M. Deitzel, J. Kleinmeyer, D. Harris and N.C. Beck Tan, *Polymer*, 42, 261 (2001).
[73] C.J. Buchko, L.C. Chen, Y. Shen and D.C. Martin, *Polymer*, 40, 7397 (1999).
[74] J.S. Lee, K.H. Choi, H.D. Ghim, S.S. Kim, D.H. Chun, H.Y. Kim and W.S. Lyoo, *J. Appl. Polym. Sci.*, 93, 1638 (2004).
[75] S.F. Fennessey and R.J. Farris, *Polymer*, 45, 4217 (2004).
[76] S. Kidoaki, I. K. Kwon and T. Matsuda, *Biomaterials*, 26, 37 (2005).
[77] X. Zong, K. Kim, D. Fang, S. Ran, B.S. Hsiao and B. Chu, *Polymer*, 43, 4403 (2002).
[78] D. Li and Y. Xia, *Nano. Lett.*, 3, 555 (2003).
[79] W.-Z. Jin, H.-W. Duan, Y.-J. Zhang and F.-F. Li, Nonafiber membrane of EVOH-based ionomer by electrospinning, *Proceedings of the 1st IEEE International Conference on Nano/Micro Engineered and Molecular Systems*, Zhuhai, China (2006).
[80] X.M. Mo, C.Y. Xu, M. Kotaki and S. Ramakrishna, *Biomaterials*, 25, 1883 (2004).
[81] S. Zhao, X. Wu, L. Wang and Y. Huang, *J. Appl. Polym. Sci.*, 91, 242 (2004).

APPENDIX

Table 8. Natural and coded variables for experimental and test data along with corresponding responses

No.	Natural Variables				Coded Variables				Responses	
	C (%)	*d* (*cm*)	*V* (*kV*)	*Q* (*ml*/*h*)	x_1	x_2	x_3	x_4	MFD (*nm*)	StdFD (*nm*)
1	8	10	15	0.2	-1	-1	-1	-1	232.62	26.60
2	8	10	15	0.3	-1	-1	-1	0	235.50	24.52
3	8	10	15	0.4	-1	-1	-1	1	252.02	25.89
4	8	10	20	0.2	-1	-1	0	-1	236.84	37.30
5	8	10	20	0.3	-1	-1	0	0	232.08	30.22
6	8	10	20	0.4	-1	-1	0	1	249.21	34.49
7	8	10	25	0.2	-1	-1	1	-1	196.05	34.76
8	8	10	25	0.3	-1	-1	1	0	201.38	35.15
9	8	10	25	0.4	-1	-1	1	1	215.00	39.00
10	8	15	15	0.2	-1	0	-1	-1	221.10	28.88
11	8	15	15	0.3	-1	0	-1	0	238.63	20.17
12	8	15	15	0.4	-1	0	-1	1	242.32	21.99
13	8	15	20	0.2	-1	0	0	-1	219.76	36.19
14	8	15	20	0.3	-1	0	0	0	228.56	28.29
15	8	15	20	0.4	-1	0	0	1	242.01	28.30
16	8	15	25	0.2	-1	0	1	-1	202.62	33.22
17	8	15	25	0.3	-1	0	1	0	208.21	37.14

Table 8. (Continued)

No.	Natural Variables				Coded Variables				Responses	
	C (%)	d (cm)	V (kV)	Q (ml/h)	x_1	x_2	x_3	x_4	MFD (nm)	StdFD (nm)
18	8	15	25	0.4	-1	0	1	1	213.66	34.84
19	8	20	15	0.2	-1	1	-1	-1	196.63	30.69
20	8	20	15	0.3	-1	1	-1	0	197.73	24.55
21	8	20	15	0.4	-1	1	-1	1	206.28	22.11
22	8	20	20	0.2	-1	1	0	-1	206.69	31.56
23	8	20	20	0.3	-1	1	0	0	224.38	27.41
24	8	20	20	0.4	-1	1	0	1	242.06	26.51
25	8	20	25	0.2	-1	1	1	-1	205.25	40.32
26	8	20	25	0.3	-1	1	1	0	215.70	30.54
27	8	20	25	0.4	-1	1	1	1	231.34	32.40
28	10	10	15	0.2	0	-1	-1	-1	269.91	30.35
29	10	10	15	0.3	0	-1	-1	0	270.05	28.88
30	10	10	15	0.4	0	-1	-1	1	291.99	33.98
31	10	10	20	0.2	0	-1	0	-1	256.11	38.54
32	10	10	20	0.3	0	-1	0	0	264.86	35.70
33	10	10	20	0.4	0	-1	0	1	278.34	49.13
34	10	10	25	0.2	0	-1	1	-1	228.21	42.33
35	10	10	25	0.3	0	-1	1	0	239.28	40.30
36	10	10	25	0.4	0	-1	1	1	238.74	46.57
37	10	15	15	0.2	0	0	-1	-1	263.67	34.16
38	10	15	15	0.3	0	0	-1	0	269.29	31.54
39	10	15	15	0.4	0	0	-1	1	277.71	29.40
40	10	15	20	0.2	0	0	0	-1	284.20	38.18
41	10	15	20	0.3	0	0	0	0	281.82	36.27
42	10	15	20	0.4	0	0	0	1	282.39	42.07
43	10	15	25	0.2	0	0	1	-1	249.42	40.79
44	10	15	25	0.3	0	0	1	0	278.22	46.15
45	10	15	25	0.4	0	0	1	1	286.96	51.16
46	10	20	15	0.2	0	1	-1	-1	239.45	27.98
47	10	20	15	0.3	0	1	-1	0	244.04	27.43
48	10	20	15	0.4	0	1	-1	1	251.58	27.26
49	10	20	20	0.2	0	1	0	-1	285.67	35.62
50	10	20	20	0.3	0	1	0	0	273.05	30.74
51	10	20	20	0.4	0	1	0	1	280.62	34.66
52	10	20	25	0.2	0	1	1	-1	278.10	40.79
53	10	20	25	0.3	0	1	1	0	280.95	44.58
54	10	20	25	0.4	0	1	1	1	306.28	44.04
55	12	10	15	0.2	1	-1	-1	-1	286.23	27.12
56	12	10	15	0.3	1	-1	-1	0	295.60	32.91
57	12	10	15	0.4	1	-1	-1	1	293.41	40.48
58	12	10	20	0.2	1	-1	0	-1	271.20	34.86
59	12	10	20	0.3	1	-1	0	0	291.89	42.78
60	12	10	20	0.4	1	-1	0	1	295.93	49.43
61	12	10	25	0.2	1	-1	1	-1	234.13	39.31
62	12	10	25	0.3	1	-1	1	0	247.65	48.60
63	12	10	25	0.4	1	-1	1	1	247.13	59.02
64	12	15	15	0.2	1	0	-1	-1	271.93	33.05
65	12	15	15	0.3	1	0	-1	0	297.65	26.75

No.	Natural Variables				Coded Variables				Responses	
	C (%)	d (cm)	V (kV)	Q (ml/h)	x_1	x_2	x_3	x_4	MFD (nm)	StdFD (nm)
66	12	15	15	0.4	1	0	-1	1	296.79	39.84
67	12	15	20	0.2	1	0	0	-1	297.94	38.82
68	12	15	20	0.3	1	0	0	0	310.06	36.84
69	12	15	20	0.4	1	0	0	1	312.15	41.69
70	12	15	25	0.2	1	0	1	-1	272.24	39.55
71	12	15	25	0.3	1	0	1	0	282.04	42.35
72	12	15	25	0.4	1	0	1	1	288.00	51.72
73	12	20	15	0.2	1	1	-1	-1	259.63	34.63
74	12	20	15	0.3	1	1	-1	0	278.40	25.35
75	12	20	15	0.4	1	1	-1	1	279.25	27.25
76	12	20	20	0.2	1	1	0	-1	307.42	42.25
77	12	20	20	0.3	1	1	0	0	327.77	35.71
78	12	20	20	0.4	1	1	0	1	337.88	45.16
79	12	20	25	0.2	1	1	1	-1	321.78	46.21
80	12	20	25	0.3	1	1	1	0	334.54	40.68
81	12	20	25	0.4	1	1	1	1	342.45	47.94
82	9	20	15	0.3	-0.5	1	-1	0	216.53	24.25
83	10	12.5	15	0.3	0	-0.5	-1	0	259.61	25.67
84	10	20	22.5	0.3	0	1	0.5	0	300.27	35.71
85	10	20	15	0.25	0	1	-1	-0.5	235.04	29.64
86	9	12.5	15	0.3	-0.5	-0.5	-1	0	247.57	26.65
87	9	20	22.5	0.3	-0.5	1	0.5	0	247.16	31.12
88	9	20	15	0.25	-0.5	1	-1	-0.5	212.82	30.26
89	10	12.5	22.5	0.3	0	-0.5	0.5	0	263.70	45.06
90	10	12.5	15	0.25	0	-0.5	-1	-0.5	258.26	26.16
91	10	20	22.5	0.25	0	1	0.5	-0.5	272.03	36.28
92	9	12.5	22.5	0.3	-0.5	-0.5	0.5	0	235.75	33.16
93	9	12.5	15	0.25	-0.5	-0.5	-1	-0.5	244.43	24.87
94	9	20	22.5	0.25	-0.5	1	0.5	-0.5	252.50	36.01
95	10	12.5	22.5	0.25	0	-0.5	0.5	-0.5	260.71	42.25
96	9	12.5	22.5	0.25	-0.5	-0.5	0.5	-0.5	231.97	32.86

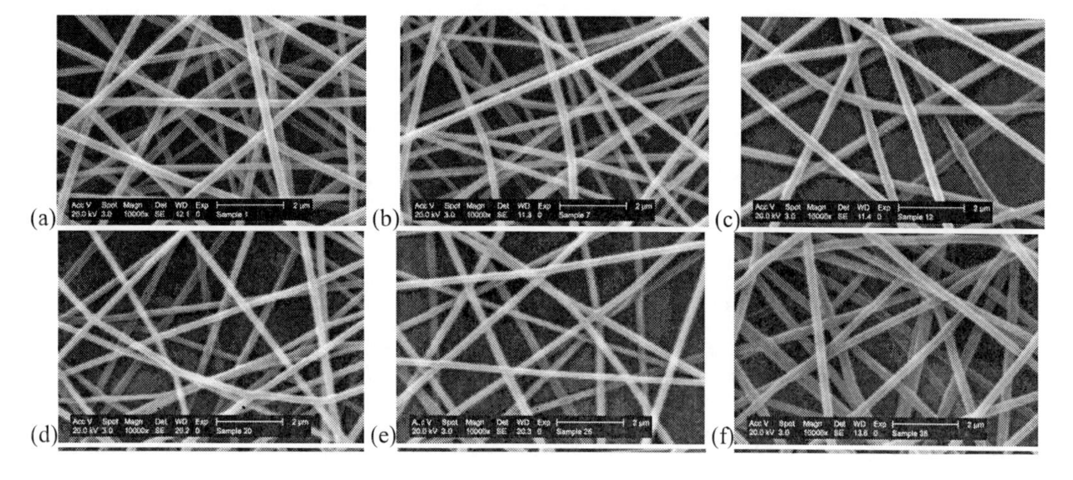

Figure 5. (Continued).

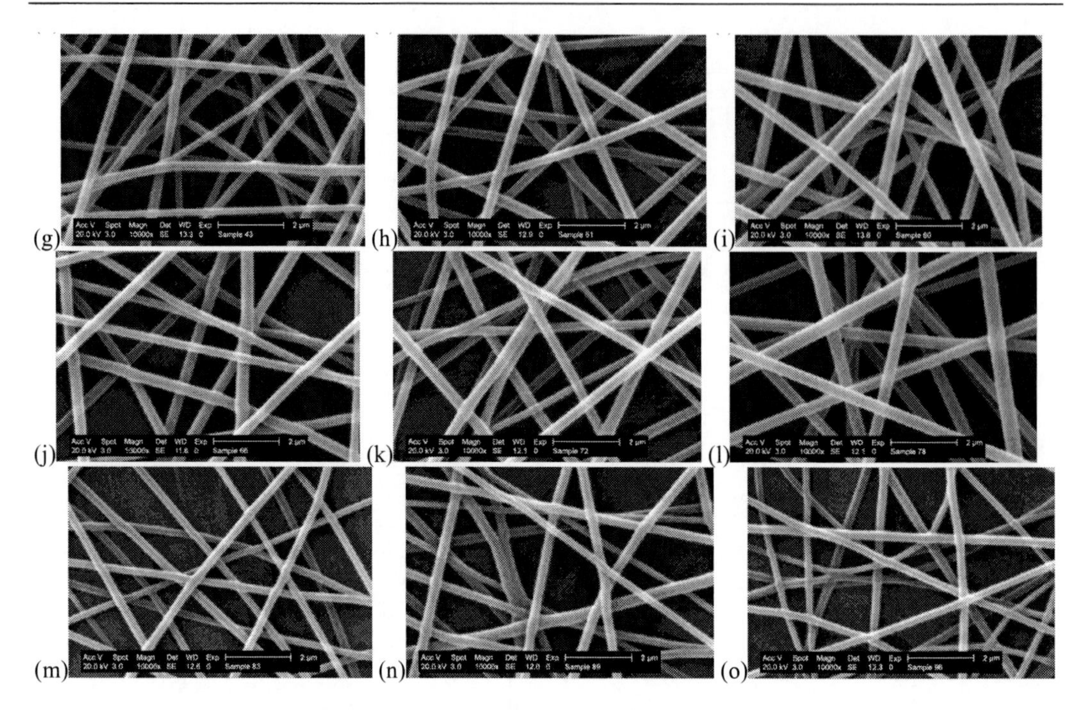

Figure 5. SEM micrographs of typical PVA electrospun nanofiber mats: (a) C= 8%, d= 10*cm*, V= 15*kV*, Q= 0.2*ml/h*, (b) C= 8%, d= 10*cm*, V= 25*kV*, Q= 0.2*ml/h*, (c) C= 8%, d= 15*cm*, V= 15*kV*, Q= 0.4*ml/h*, (d) C= 8%, d= 20*cm*, V= 15*kV*, Q= 0.3*ml/h*, (e) C= 8%, d= 20*cm*, V= 25*kV*, Q= 0.3*ml/h*, (f) C= 10%, d= 15*cm*, V= 15*kV*, Q= 0.3*ml/h*, (g) C= 10%, d= 15*cm*, V= 25*kV*, Q= 0.2*ml/h*, (h) C= 10%, d= 20*cm*, V= 20*kV*, Q= 0.4*ml/h*, (i) C= 12%, d= 10*cm*, V= 20*kV*, Q= 0.4*ml/h*, (j) C= 12%, d= 15*cm*, V= 15*kV*, Q= 0.4*ml/h*, (k) C= 12%, d= 15*cm*, V= 25*kV*, Q= 0.4*ml/h*, (l) C= 12%, d= 20*cm*, V= 20*kV*, Q= 0.4*ml/h*, (m) C= 10%, d= 12.5*cm*, V= 15*kV*, Q= 0.3*ml/h*, (n) C= 10%, d= 12.5*cm*, V= 22.5*kV*, Q= 0.3*ml/h*, (o) C= 9%, d= 12.5*cm*, V= 22.5*kV*, Q= 0.25*ml/h*.

Chapter 2

LAMINATED NANOCOMPOSITES

1. INTRODUCTION

Polymeric nanofibers can be made using the electrospinning process, has already been described in the literature [1-10]. Electrospinning (Figure 1) uses a high electric field to draw a polymer solution from tip of a capillary toward a collector [16-20]. A voltage is applied to the polymer solution, which causes a jet of the solution to be drawn toward a grounded collector. The fine jets dry to form polymeric fibers, which can be collected as a web [11-17].

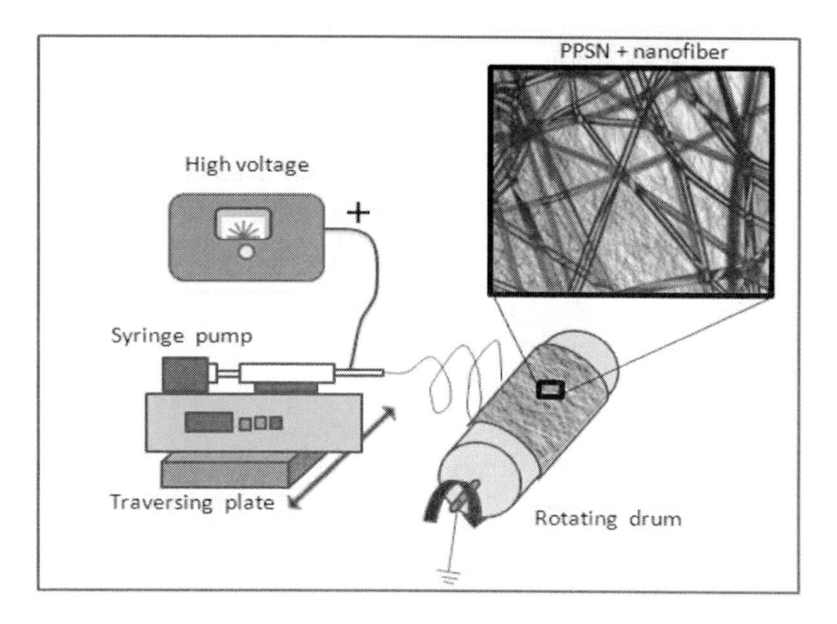

Figure 1. Ecectrosprinning setup.

In the non woven industry on of the fastest growing segments is in filtration applications. Traditionally wet-laid, melt blown and spun non woven articles, containing micron size fibers are most popular for these applications because of the low cost, easy process ability and good filtration efficiency. Their applications in filtration can be divided into two major areas: air filtration and liquid filtration (Figure 2).

Another type of electrospinning equipment (Figure 2) also used a variable high voltage power supply from Gamma High Voltage Research (USA). The applied voltage can be varied from 1- 30 kV. A 5-ml syringe was used and positive potential was applied to the polymer blend solution by attaching the electrode directly to the outside of the hypodermic needle with internal diameter of 0.3 mm. The collector screen was a 20×20 cm aluminum foil, which was placed 10 cm horizontally from the tip of the needle. The electrode of opposite polarity was attached to the collector. A metering syringe pump from New Era pump systems Inc. (USA) was used. It was responsible for supplying polymer solution with a constant rate of 20 μl/min.

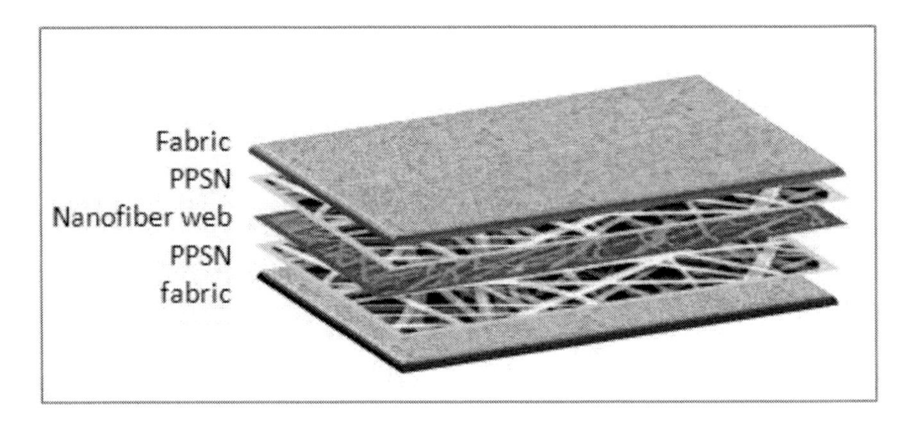

Figure 2. Multilayer fabric components.

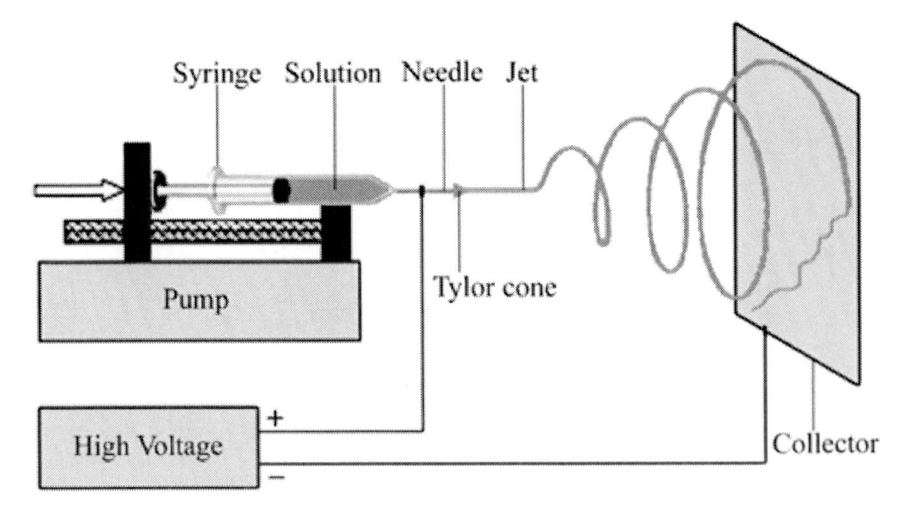

Figure 3. Schematic diagram of general type of electrospinning apparatus.

Electrospinning was done in a temperature-controlled chamber and temperature of electrospinning environment was adjusted on variable temperatures. Schematic diagram of the electrospinning apparatus was shown in Figure 3.

Elecrospinning is a process that produces continuous polymer fibers with diameter in the submicron range. In the electrospinnig process the electric body force act on element of charged fluid. Electrospinning has emerged as a specialized processing technique for the formation of sub-micron fibers (typically between 100 nm and 1 μm in diameter), with high specific surface areas. Due to their high specific surface area, high porosity, and small pore size, the unique fibers have been suggested for excellent condidate for use in filtration [18-22].

Air and water are the bulk transportation medium for transmission of particulate contaminants. The contaminants during air filtration are complex mixture of particles. The most of them are usually smaller than 1000 μm in diameter chemical and biological aerosols are frequently in range of 1-10 μm. The particulate matters may carry some gaseous contaminants. In water filtration removal of particulate and biological contaminants is an important step. Nowadays, the filtration industry is looking for energy efficient high performance filters for filtration of particles smaller than 0.3 μm and adsorbed toxic gases [23-30].

Nanofibrous media have low basis weight, high permeability, and small pore size that make them appropriate for a wide range of filtration applications. In addition, nanofiber membrane offers unique properties like high specific surface area (ranging from 1to 35m2/g depending on the diameter of fibers), good interconnectivity of pores and potential to incorporate active chemistry or functionality on nanoscale. In our study, to prepare the filters, a flow rate 1 μl/h for solution was selected and the fibers were collected on an aluminum-covered rotating drum (with speed 9 m/min) which was previously covered with a polypropylene spun-bond nonwoven (PPSN) substrate of 28cm× 28cm dimensions; 0.19 mm thickness; 25 g/m2 weight; 824 $cm^3/s/cm^2$ air permeability and 140°C melting point (Figure 2).

Structure characteristics of nanofiberous filtering media such as layer thickness ,fiber diameter,nano fiber orientation, representative pore size, porosity dictate the filter properties and quality.

Clearly, the properties of a nanofiberious media will depend on its structural charcteristics as well as the nature of the component fibers.thus it is desirable to understand and determine these charactristics.In this work tryed to identify the orientation distribution function (ODF) of nanofibers in nanofilter , the fiber thickness distribution and porosity of nanofiberious media by using image processing algoritms.

– Effect of systematic parameters on electrospun nanofibers

It has been found that morphology such as fiber diameter and its uniformity of the electrospun nano fibers are dependent on many processing parameters. These parameters can be divided into three main groups: a) solution properties, b) processing conditions, c) ambient conditions. Each of the parameters has been found to affect the morphology of the electrospun fibers.

– Solution Properties

Parameters such as viscosity of solution, solution concentration, molecular weight of solution, electrical conductivity, elasticity and surface tension, have important effect on morphology of nanofibers.

– Viscosity

The viscosity range of a different nanofiber solution which is spinnable is different. One of the most significant parameters influencing the fiber diameter is the solution viscosity. A higher viscosity results in a large fiber diameter. Beads and beaded fibers are less likely to be formed for the more viscous solutions. The diameter of the beads become bigger and the average distance between beads on the fibers longer as the viscosity increases.

– Solution concentration

In electrospinning process, for fiber formation to occur, a minimum solution concentration is required. As the solution concentration increase, a mixture of beads and fibers is obtained. The shape of the beads changes from spherical to spindle-like when the solution concentration varies from low to high levels. It should be noted that the fiber diameter increases with increasing solution concentration because the higher viscosity resistance. Nevertheless, at higher concentration, viscoelastic force which usually resists rapid changes in fiber shape may result in uniform fiber formation. However, it is impossible to electrospin if the solution concentration or the corresponding viscosity become too high due to the difficulty in liquid jet formation.

– Molecular weight

Molecular weight also has a significant effect on the rheological and electrical properties such as viscosity, surface tension, conductivity and dielectric strength. It has been observed that too low molecular weight solution tend to form beads rather than fibers and high molecular weight nanofiber solution give fibers with larger average diameter.

– Surface tension

The surface tension of a liquid is often defined as the force acting at right angles to any line of unit length on the liquid surface. By reducing surface tension of a nanofiber solution, fibers could be obtained without beads. The surface tension seems more likely to be a function of solvent compositions, but is negligibly dependent on the solution concentration. Different solvents may contribute different surface tensions. However, not necessarily a lower surface tension of a solvent will always be more suitable for electrospinning. Generally, surface tension determines the upper and lower boundaries of electrospinning window if all other variables are held constant. The formation of droplets, bead and fibers can be driven by the surface tension of solution and lower surface tension of the spinning solution helps electrospinning to occur at lower electric field.

– Solution conductivity

There is a significant drop in the diameter of the electrospun nanofibers when the electrical conductivity of the solution increases. Beads may also be observed due to low conductivity of the solution, which results in insufficient elongation of a jet by electrical force to produce uniform fiber. In general, electrospun nanofibers with the smallest fiber diameter can be obtained with the highest electrical conductivity. This interprets that the drop in the size of the fibers is due to the increased electrical conductivity.

– Applied voltage

In the case of electrospinning, the electric current due to the ionic conduction of charge in the nanofiber solution is usually assumed small enough to be negligible. The only mechanism of charge transport is the flow of solution from the tip to the target. Thus, an increase in the electrospinning current generally reflects an increase in the mass flow rate from the capillary tip to the grounded target when all other variables (conductivity, dielectric constant, and flow rate of solution to the capillary tip) are held constant.Increasing the applied voltage (*i.e.*, increasing the electric field strength) will increase the electrostatic repulsive force on the fluid jet which favors the thinner fiber formation. On the other hand, the solution will be removed from the capillary tip more quickly as jet is ejected from Taylor cone. This results in the increase of the fiber diameter.

– Feed Rate

The morphological structure can be slightly changed by changing the solution flow rate. When the flow rate exceeded a critical value, the delivery rate of the solution jet to the capillary tip exceeds the rate at which the solution was removed from the tip by the electric forces. This shift in the mass-balance resulted in sustained but unstable jet and fibers with big beads formation.

In the first part of this study, the production of electrospun nanofibers investigated. In another part, a different case study presented to show how nanofibers can be laminated for application in filter media.

2. Experiments: Case1-Production of Nanofibers

2.1. Preparation of Regenerated SF Solution

Raw silk fibers (B.mori cocoons were obtained from domestic producer, Abrisham Guilan Co., IRAN) were degummed with 2 gr/L Na_2CO_3 solution and 10 gr/L anionic detergent at 100 ° C for 1 h and then rinsed with warm distilled water. Degummed silk (SF) was dissolved in a ternary solvent system of $CaCl_2/CH_3CH_2OH/H_2O$ (1:2:8 in molar ratio) at 70 ° C for 6 h. After dialysis with cellulose tubular membrane (Bialysis Tubing D9527 Sigma) in H_2O for 3 days, the SF solution was filtered and lyophilized to obtain the regenerated SF sponges.

2.2. Preparation of the Spinning Solution

SF solutions were prepared by dissolving the regenerated SF sponges in 98% formic acid for 30 min. Concentrations of SF solutions for electrospinning was in the range from 8% to 14% by weight.

2.3. Electrospinning

In the electrospinning process, a high electric potential (Gamma High voltage) was applied to a droplet of SF solution at the tip (0.35 mm inner diameter) of a syringe needle, The electrospun nanofibers were collected on a target plate which was placed at a distance of 10 cm from the syringe tip. The syringe tip and the target plate were enclosed in a chamber for adjusting and controlling the temperature. Schematic diagram of the electrospinning apparatus is shown in Figure 2. The processing temperature was adjusted at 25, 50 and 75 °C. A high voltage in the range from 10 kV to 20 kV was applied to the droplet of SF solution.

2.4. Characterization

Optical microscope (Nikon Microphot-FXA) was used to investigate the macroscopic morphology of electrospun SF fibers. For better resolving power, morphology, surface texture and dimensions of the gold-sputtered electrospun nanofibers were determined using a Philips XL-30 scanning electron microscope. A measurement of about 100 random fibers was used to determine average fiber diameter and their distribution.

3. Experiment: Case2-Production of Laminated Composites

Polyacrylonitrile (PAN) of 70,000 g/mol molecular weight from Polyacryl Co. (Isfehan, Iran) has been used with Dimethylformamide (DMF) from Merck, to form a polymer solution 12% w/w after stirring for 5 h and staying overnight under room temperature. The yellow and ripen solution was inserted into a plastic syringe with a stainless steel nozzle 0.4 mm in inner diameter and then it was placed in a metering pump from WORLD PRECISION INSTRUMENTS (Florida, USA). Next, this set installed on a plate which it could traverse to left-right along drum (Fig.1). The flow rate 1 μl/h for solution was selected and the fibers were collected on an aluminum-covered rotating drum (with speed 9 m/min) which was previously covered with a polypropylene spun-bond nonwoven (PPSN) substrate of 28cm× 28cm dimensions; 0.19 mm thickness; 25 g/m2 weight; 824 $cm^3/s/cm^2$ air permeability and 140°C melting point. The distance between the nozzle and the drum was 7cm and an electric voltage of approximately 11kV was applied between them. Eelectrospinning process was carried out for 8h at room temperature to reach approximately web thickness 3.82 g/m^2. Then nanofiber webs were laminated into cotton weft-warp fabric with a thickness 0.24mm and density of 25×25 (warp-weft) per centimeter to form a multilayer fabric (Fig. 2). Laminating was performed at temperatures 85,110,120,140,160°C for 1 min under a pressure of 9 gf/cm^2.

Air permeability of multilayer fabric before and after lamination was tested by TEXTEST FX3300 instrument (Zürich, Switzerland). Also, in order to consider of nanofiber morphology after hot-pressing, another laminating was performed by a non-stick sheet made of Teflon (0.25 mm thickness) instead one of the fabrics (fabric /pp web/nanofiber web/pp web/non-stick sheet). Finally, after removing of Teflon sheet, the nanofiber layer side was observed under an optical microscope (MICROPHOT-FXA, Nikon, Japan) connected to a digital camera.

4. Results and Discussion

4.1. Effect of Silk Concentration

One of the most important quantities related with electrospun nanofibers is their diameter. Since nanofibers are resulted from evaporation of polymer jets, the fiber diameters will depend on the jet sizes and the solution concentration. It has been reported that during the traveling of a polymer jet from the syringe tip to the collector, the primary jet may be split into different sizes multiple jets, resulting in different fiber diameters. When no splitting is involved in electrospinning, one of the most important parameters influencing the fiber diameter is Concentration of regenerated silk solution. The jet with a low concentration breaks into droplets readily and a mixture of fibers, bead fibers and droplets as a result of low viscosity is generated. These fibers have an irregular morphology with large variation in size, on the other hand jet with high concentration don't break up but traveled to the grounded target and tend to facilitate the formation of fibers without beads and droplets. In this case, Fibers became more uniform with regular morphology.

At first, a series of experiments were carried out when the silk concentration was varied from 8 to 14% at the 15KV constant electric field and 25 ° C constant temperature. Below the silk concentration of 8% as well as at low electric filed in the case of 8% solution, droplets were formed instead of fibers. Fig. 2 shows morphology of the obtained fibers from 8% silk solution at 20 KV. The obtained fibers are not uniform. The average fiber diameter is 72 nm and a narrow distribution of fiber diameters is observed. It was found that continues nanofibers were formed above silk concentration of 8% regardless of the applied electric field and electrospinning condition. In the electrospinning of silk fibroin, when the silk concentration is more than 10%, thin and rod like fibers with diameters range from 60-450 nm were obtained.

There is a significant increase in mean fiber diameter with the increasing of the silk concentration, which shows the important role of silk concentration in fiber formation during electrospinning process. Concentration of the polymer solution reflects the number of entanglements of polymer chains in the solution, thus solution viscosity. Experimental observations in electrospinning confirm that for forming fibers, a minimum polymer concentration is required. Below this critical concentration, application of electric field to a polymer solution results electrospraying and formation of droplets to the instability of the ejected jet. As the polymer concentration increased, a mixture of beads and fibers is formed.

Further increase in concentration results in formation of continuous fibers as reported in this paper. It seems that the critical concentration of the silk solution in formic acid for the formation of continuous silk fibers is 10%.

Experimental results in electrospinning showed that with increasing the temperature of electrospinning process, concentration of polymer solution has the same effect on fibers diameter at 25 °C.

There is a significant increase in mean fiber diameter with increasing of the silk concentration, which shows the important role of silk concentration in fiber formation during electrospinning process. It is well known that the viscosity of polymer solutions is proportional to concentration and polymer molecular weight. For concentrated polymer solution, concentration of the polymer solution reflects the number of entanglements of polymer chains, thus have considerable effects on the solution viscosity. At fixed polymer molecular weight, the higher polymer concentration resulting higher solution viscosity. The jet from low viscosity liquids breaks up into droplets more readily and few fibers are formed, while at high viscosity, electrospinning is prohibit because of the instability flow causes by the high cohesiveness of the solution. Experimental observations in electrospinning confirm that for fiber formation to occur, a minimum polymer concentration is required. Below this critical concentration, application of electric field to a polymer solution results electro spraying and formation of droplets to the instability of the ejected jet. As the polymer concentration increased, a mixture of beads and fibers is formed. Further increase in concentration results in formation of continuous fibers as reported in this chapter. It seems that the critical concentration of the silk solution in formic acid for the formation of continuous silk fibers is 10% when the applied electric field was in the range of 10 to 20 kV.

4.2. Effect of Electric Field

It was already reported that the effect of the applied electrospinning voltage is much lower than effect of the solution concentration on the diameter of electrospun fibers. In order to study the effect of the electric field, silk solution with the concentration of 10%, 12%, and 14% were electrospun at 10, 15, and 20 KV at 25 °C. At a high solution concentration, Effect of applied voltage is nearly significant. It is suggested that, at this temperature, higher applied voltage causes multiple jets formation, which would provide decrees fiber diameter.

As the results of this finding it seems that electric field shows different effects on the nanofibers morphology. This effect depends on the polymer solution concentration and electrospinning conditions.

4.3. Effect of Electrospinning Temperature

One of the most important quantities related with electrospun nanofibers is their diameter. Since nanofibers are resulted from evaporation of polymer jets, the fiber diameters will depend on the jet sizes. The elongation of the jet and the evaporation of the solvent both change the shape and the charge per unit area carried by the jet. After the skin is formed, the solvent inside the jet escapes and the atmospheric pressure tends to collapse the tube like jet. The circular cross section becomes elliptical and then flat, forming a ribbon-like structure. In this work we believe that ribbon-like structure in the electrospinning of SF at higher temperature thought to be related with skin formation at the jets. With increasing the electrospinning temperature, solvent evaporation rate increases, which results in the formation of skin at the jet surface. Non- uniform lateral stresses around the fiber due to the uneven evaporation of solvent and/or striking the target make the nanofibers with circular cross-section to collapse into ribbon shape.

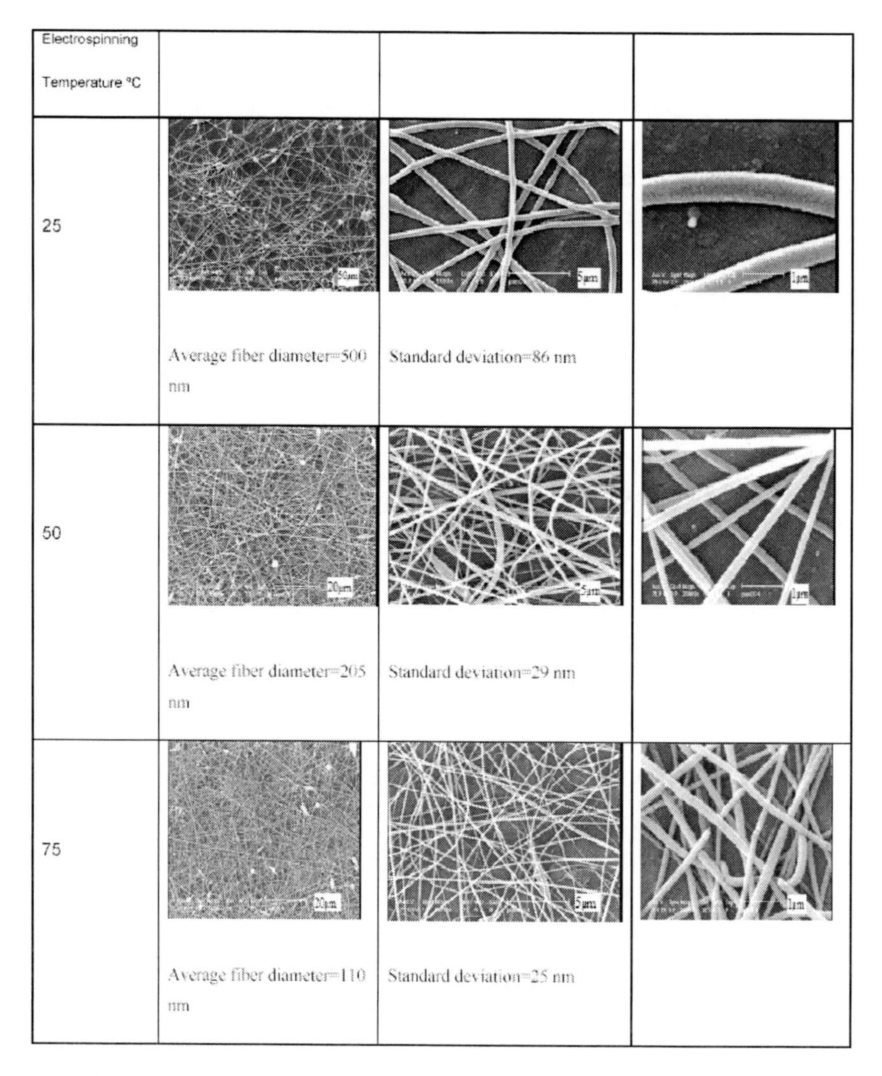

Figure 4. SEM micrographs of electrospun nanofibers at applied voltage of 20 kV and PANI content of 20% with a constant spinning distance of 10cm.

Bending of the electrospun ribbons were observed on the SEM micrographs as a result of the electrically driven bending instability or forces that occurred when the ribbon was stopped on the collector. Another problem that may be occurring in the electrospinning of SF at high temperature is the branching of jets. With increasing the temperature of electrospinning process, the balance between the surface tension and electrical forces can shift so that the shape of a jet becomes unstable. Such an unstable jet can reduce its local charge per unit surface area by ejecting a smaller jet from the surface of the primary jet or by splitting apart into two smaller jets. Branched jets, resulting from the ejection of the smaller jet on the surface of the primary jet were observed in electrospun fibers of SF. The axes of the cones from which the secondary jets originated were at an angle near 90° with respect to the axis of the primary jet.

In order to study the effect of electrospinning temperature on the morphology and texture of electrospun silk nanofibers, 12% silk solution was electrospun at various temperatures of 25, 50 and 75 °C. Results are shown in Fig. 15. Interestingly, the electrospinning of silk solution showed flat fiber morphology at 50 and 75 °C, whereas circular structure was observed at 25 °C. At 25 °C, the nanofibers with a rounded cross section and a smooth surface were collected on the target. Their diameter showed a size range of approximately 100 to 300 nm with 180 nm being the most frequently occurring. They are within the same range of reported size for electrospun silk nanofibers. With increasing the electrospinning temperature to 50 °C, The morphology of the fibers was slightly changed from circular cross section to ribbon like fibers. Fiber diameter was also increased to a range of approximately 20 to 320 nm with 180 nm the most occurring frequency. At 75 °C, The morphology of the fibers was completely changed to ribbon like structure. Furthermore, fibers dimensions were increased significantly to the range of 500 to 4100 nm with 1100 nm the most occurring frequency. The results are shown in Figure 4.

5. Experimental Design

Response surface methodology (RSM) is a collection of mathematical and statistical techniques for empirical model building (Appendix). By careful design of *experiments*, the objective is to optimize a *response* (output variable) which is influenced by several *independent variables* (input variables). An experiment is a series of tests, called *runs*, in which changes are made in the input variables in order to identify the reasons for changes in the output response.

In order to optimize and predict the morphology and average fiber diameter of electrospun silk, design of experiment was employed in the present work. Morphology of fibers and distribution of fiber diameter of silk precursor were investigated varying concentration, temperature and applied voltage. A more systematic understanding of these process conditions was obtained and a quantitative basis for the relationships between average fiber diameter and electrospinning parameters was established using response surface methodology (Appendix), which will provide some basis for the preparation of silk nanofibers.

A central composite design was employed to fit a second-order model for three variables. Silk concentration (X_1), applied voltage (X_2), and temperature (X_3) were three independent variables (factors) considered in the preparation of silk nanofibers, while the fibers diameter were dependent variables (response). The actual and corresponding coded values of three factors (X_1, X_2, and X_3) are given in Table 1. The following second-order model in X_1, X_2 and X_3 was fitted using the data in Table 1:

$$Y = \beta_0 + \beta_1 x_1 + \beta_2 x_2 + \beta_3 x_3 + \beta_{11} x_1^2 + \beta_{22} x_2^2 + \beta_{33} x_3^2 + \beta_{12} x_1 x_2 + \beta_{13} x_1 x_3 + \beta_{23} x_2 x_3 + \varepsilon$$

Table 1. Central composite design

X_i	Independent variables	Coded values -1	0	1
X_1	Silk concentration (%)	10	12	14
X_2	applied voltage (KV)	10	15	20
X_3	temperature (° C)	25	50	75

The Minitab and Mathlab programs were used for analysis of this second-order model and for response surface plots (Minitab 11, Mathlab 7).

By Regression analysis, values for coefficients for parameters and P-values (a measure of the statistical significance) are calculated. When P-value is less than 0.05, the factor has significant impact on the average fiber diameter. If P-value is greater than 0.05, the factor has no significant impact on average fiber diameter. And R^2_{adj} (represents the proportion of the total variability that has been explained by the regression model) for regression models were obtained (Table 2). The fitted second-order equation for average fiber diameter can be considered by:

$$Y = 391 + 311 X_1 - 164 X_2 + 57 X_3 - 162 X_1^2 + 69 X_2^2 + 391 X_3^2 - 159 X_1X_2 + 315 X_1X_3 - 144 X_2X_3 \quad (1)$$

Where Y = Average fiber diameter

Table 2. Regression Analysis for the three factors (concentration, applied voltage, temperature) and coefficients of the model in coded unit*

Variables	Constant		P-value
	β_0	391.3	0.008
x_1	β_1	310.98	0.00
x_2	β_2	-164.0	0.015
x_3	β_3	57.03	0.00
x^2_1	β_{11}	161.8	0.143
x^2_2	β_{22}	68.8	0.516
x^2_3	β_{33}	390.9	0.002
x_1x_2	β_{12}	-158.77	0.048
x_1x_3	β_{13}	314.59	0.001
x_2x_3	β_{23}	-144.41	0.069
F	P-value	R^2	R^2 (adj)
18.84	0.00	0.907	0.858

* Model: $Y=\beta o+\beta_1x_1+\beta_2x_2+\beta_3x_3+\beta_{11}x^2_1+\beta_{22}x^2_2+\beta_{332}x^2_3+\beta_{12}x_1x2+\beta_{13}x_1x\ 3+\beta_{13}x_2x3$ where "y" is average fiber diameter.

From the P-values listed in Table 2, it is obvious that P-value of term X_2 is greater than P-values for terms X_1 and X_3. And other P-values for terms related to applied voltage such as, X_2^2, X_1X_2, X_2X_3 are much greater than significance level of 0.05. That is to say, applied voltage has no much significant impact on average fiber diameter and the interactions between concentration and applied voltage, temperature and applied voltage are not significant, either. But P-values for term related to X_3 and X_1 are less than 0.05. Therefore, temperature and concentration have significant impact on average fiber diameter. Furthermore, R^2_{adj} is 0.858, That is to say, this model explains 86% of the variability in new data.

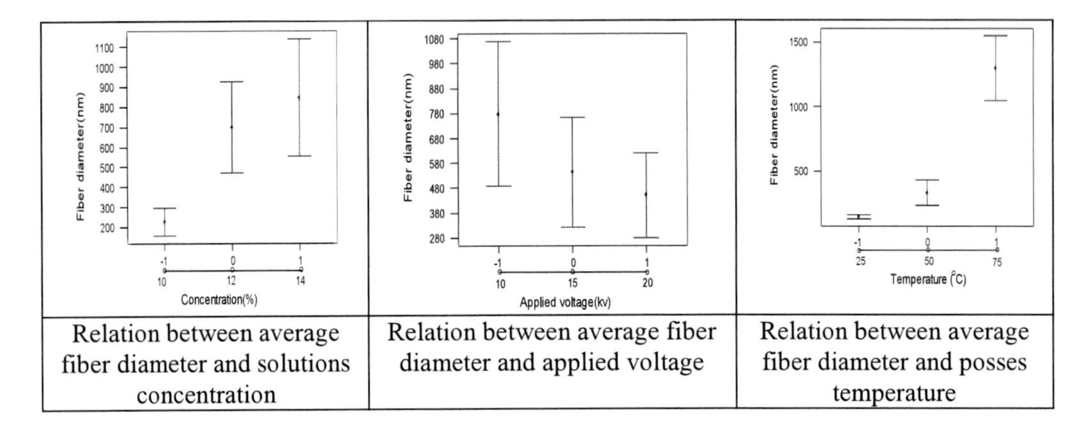

Relation between average fiber diameter and solutions concentration	Relation between average fiber diameter and applied voltage	Relation between average fiber diameter and posses temperature

Figure 5. Effect of electrospinning parameters on nanofibers diameter

CONCLUSION

In the first parts of this study, the electrospinning of silk fibroin was processed and the average fiber diameters depend on the electrospinning condition. Morphology of fibers and distribution of diameter were investigated at various concentrations, applied voltages and temperature. The electrospinning temperature and the solution concentration have a significant effect on the morphology of the electrospun silk nanofibers. There effects were explained to be due to the change in the rate of skin formation and the evaporation rate of solvents. To determine the exact mechanism of the conversion of polymer into nanofibers require further theoretical and experimental work.

From the practical view the results of the present work can be condensed. Concentration of regenerated silk solution was the most dominant parameter to produce uniform and continuous fibers. The jet with a low concentration breaks into droplets readily and a mixture of fibers and droplets as a result of low viscosity is generated. On the other hand jets with high concentration do not break up but traveled to the target and tend to facilitate the formation of fibers without beads and droplets. In this case, fibers become more uniform with regular morphology. In the electrospinning of silk fibroin, when the silk concentration is more than 10%, thin and rod like fibers with diameters range from 60-450 nm were obtained. Furthermore, In the electrospinning of silk fibroin, when the process temperature is more than 25 °C, flat, ribbon like and branched fibers with diameters range from 60-7000 nm were obtained.

Two- way analysis of variance was carried out at the significant level of 0.05 to study the impact of concentration, applied voltages and temperature on average fiber diameter. It was concluded that concentration of solution and electrospinning temperature were the most significant factors impacting the diameter of fibers. Applied voltage had no significant impact on average fiber diameter.

In the second part of this study the effect of laminating temperature on nanofiber/laminate properties is discussed. This laminating temperature is an important parameter to make next-generation filter media. Figure 6 shows the optical microscope images of nanofiber web after lamination at various temperatures. We will discuss the laminating procedure more in detail in our future publications.

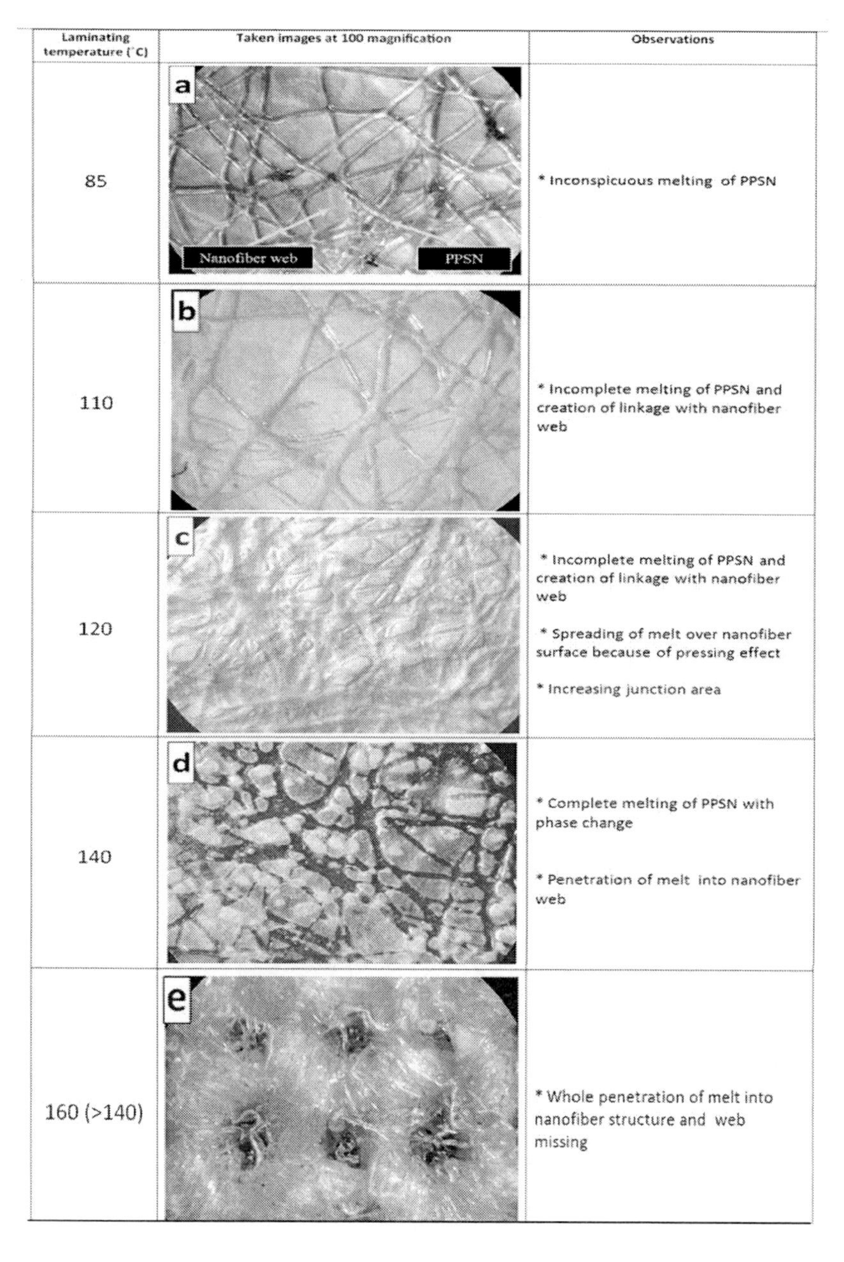

Figure 6. The optical microscope images of nanofiber web after lamination at various temperatures to be used as filter media.

APPENDIX

Variables which potentially can alter the electrospinning process (Figure A-1) are large. Hence, investigating all of them in the framework of one single research would almost be impossible. However, some of these parameters can be held constant during experimentation. For instance, performing the experiments in a controlled environmental condition, which is concerned in this study, the ambient parameters (i.e. temperature, air pressure, and humidity) are kept unchanged. Solution viscosity is affected by polymer molecular weight, solution concentration, and temperature. For a particular polymer (constant molecular weight) at a fixed temperature, solution concentration would be the only factor influencing the viscosity. In this circumstance, the effect of viscosity could be determined by the solution concentration. Therefore, there would be no need for viscosity to be considered as a separate parameter.

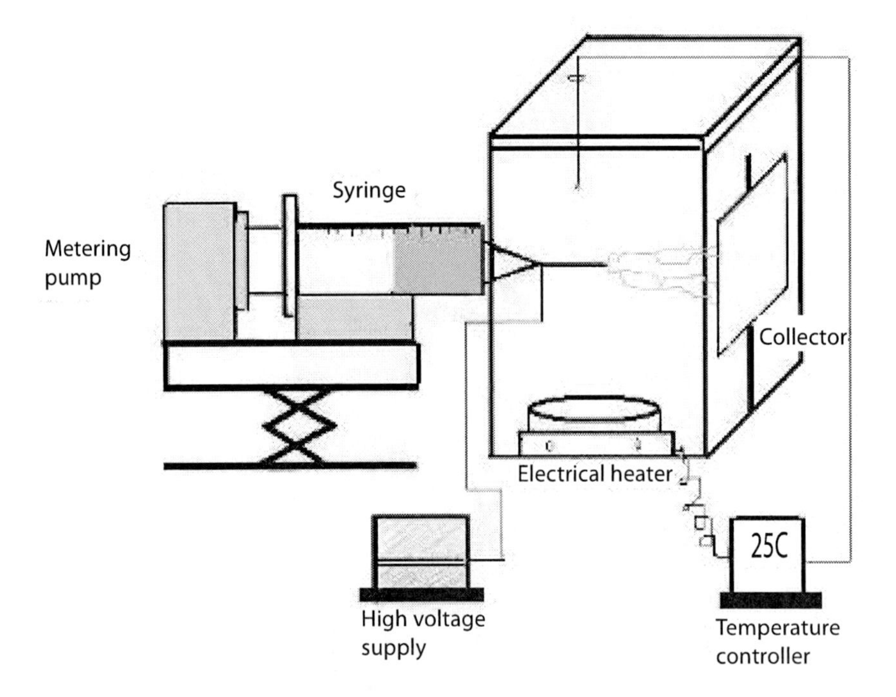

Figure A-2. A typical image of Electrospinning process

In this regard, solution concentration (C), spinning distance (d), applied voltage (V), and volume flow rate (Q) were selected to be the most influential parameters. The next step is to choose the ranges over which these factors are varied. Process knowledge, which is a combination of practical experience and theoretical understanding, is required to fulfill this step. The aim is here to find an appropriate range for each parameter where dry, bead-free, stable, and continuous fibers without breaking up to droplets are obtained. This goal could be achieved by conducting a set of preliminary experiments while having the previous works in mind along with utilizing the reported relationships.

The relationship between intrinsic viscosity ($[\eta]$) and molecular weight (M) is given by the well-known Mark-Houwink-Sakurada equation as follows:

$$[\eta] = KM^{a} \qquad \text{(A-1)}$$

where K and a are constants for a particular polymer-solvent pair at a given temperature. Polymer chain entanglements in a solution can be expressed in terms of Berry number (B), which is a dimensionless parameter and defined as the product of intrinsic viscosity and polymer concentration ($B=[\eta]C$). For each molecular weight, there is a lower critical concentration at which the polymer solution cannot be electrospun.

As for determining the appropriate range of applied voltage, referring to previous works, it was observed that the changes of voltage lay between 5 kV to 25 kV depending on experimental conditions; voltages above 25 kV were rarely used. Afterwards, a series of experiments were carried out to obtain the desired voltage domain. At $V<10\ kV$, the voltage was too low to spin fibers and $10\ kV \leq V < 15\ kV$ resulted in formation of fibers and droplets; in addition, electrospinning was impeded at high concentrations. In this regard, $15\ kV \leq V \leq 25\ kV$ was selected to be the desired domain for applied voltage.

The use of 5 cm – 20 cm for spinning distance was reported in the literature. Short distances are suitable for highly evaporative solvents whereas it results in wet coagulated fibers for nonvolatile solvents due to insufficient evaporation time. Afterwards, this was proved by experimental observations and $10\ cm \leq d \leq 20\ cm$ was considered as the effective range for spinning distance.

Few researchers have addressed the effect of volume flow rate. Therefore in this case, the attention was focused on experimental observations. At $Q<0.2\ ml/h$, in most cases especially at high polymer concentrations, the fiber formation was hindered due to insufficient supply of solution to the tip of the syringe needle. Whereas, excessive feed of solution at $Q>0.4\ ml/h$ incurred formation of droplets along with fibers. As a result, $0.2\ ml/h \leq Q \leq 0.4\ ml/h$ was chosen as the favorable range of flow rate in this study.

Consider a process in which several factors affect a response of the system. In this case, a conventional strategy of experimentation, which is extensively used in practice, is the *one-factor-at-a-time* approach. The major disadvantage of this approach is its failure to consider any possible interaction between the factors, say the failure of one factor to produce the same effect on the response at different levels of another factor. For instance, suppose that two factors A and B affect a response. At one level of A, increasing B causes the response to increase, while at the other level of A, the effect of B totally reverses and the response decreases with increasing B. As interactions exist between electrospinning parameters, this approach may not be an appropriate choice for the case of the present work. The correct strategy to deal with several factors is to use a full factorial design. In this method, factors are all varied together; therefore all possible combinations of the levels of the factors are investigated. This approach is very efficient, makes the most use of the experimental data and takes into account the interactions between factors.

It is trivial that in order to draw a line at least two points and for a quadratic curve at least three points are required. Hence, three levels were selected for each parameter in this study so that it would be possible to use quadratic models. These levels were chosen equally spaced. A full factorial experimental design with four factors (solution concentration, spinning distance, applied voltage, and flow rate) each at three levels (3^4 design) were employed resulting in 81 treatment combinations. This design is shown in Figure A-3.

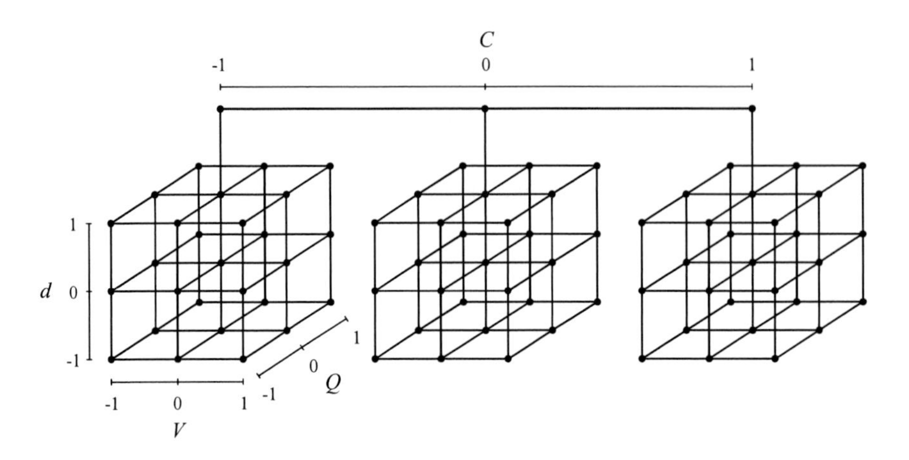

Figure A-3. 3^4 full factorial experimental design used in this study

-1, 0, and 1 are coded variables corresponding to low, intermediate and high levels of each factor respectively. The coded variables (x_j) were calculated using Equation (2) from natural variables (ξ_i). The indices 1 to 4 represent solution concentration, spinning distance, applied voltage, and flow rate respectively. In addition to experimental data, 15 treatments inside the design space were selected as test data and used for evaluation of the models. The natural and coded variables for experimental data (numbers 1-81) as well as test data (numbers 82-96) are listed in Table 8 in Appendix.

$$x_j = \frac{\xi_j - [\xi_{hj} + \xi_{lj}]/2}{[\xi_{hj} - \xi_{lj}]/2} \quad \text{(A-2)}$$

The mechanism of some scientific phenomena has been well understood and models depicting the physical behavior of the system have been drawn in the form of mathematical relationships. However, there are numerous processes at the moment which have not been sufficiently been understood to permit the theoretical approach. Response surface methodology (RSM) is a combination of mathematical and statistical techniques useful for empirical modeling and analysis of such systems. The application of RSM is in situations where several input variables are potentially influence some performance measure or quality characteristic of the process – often called responses. The relationship between the response (y) and k input variables ($\xi_1,\xi_2,...,\xi_k$) could be expressed in terms of mathematical notations as follows:

$$y = f(\xi_1, \xi_2, ..., \xi_k) \quad \text{(A-3)}$$

where the true response function f is unknown. It is often convenient to use coded variables ($x_1,x_2,..,x_k$) instead of natural (input) variables. The response function will then be:

$$y = f(x_1, x_2, ..., x_k) \quad \text{(A-4)}$$

Since the form of true response function f is unknown, it must be approximated. Therefore, the successful use of RSM is critically dependent upon the choice of appropriate function to approximate f. Low-order polynomials are widely used as approximating functions. First order (linear) models are unable to capture the interaction between parameters which is a form of curvature in the true response function. Second order (quadratic) models will be likely to perform well in these circumstances. In general, the quadratic model is in the form of:

$$y = \beta_0 + \sum_{j=1}^{k} \beta_j x_j + \sum_{j=1}^{k} \beta_{jj} x_j^2 + \sum_{i<j} \sum_{j=2}^{k} \beta_{ij} x_i x_j + \varepsilon \qquad \text{(A-5)}$$

where ε is the error term in the model. The use of polynomials of higher order is also possible but infrequent. The βs are a set of unknown coefficients needed to be estimated. In order to do that, the first step is to make some observations on the system being studied. The model in Equation (5) may now be written in matrix notations as:

$$\mathbf{y} = \mathbf{X\beta} + \boldsymbol{\varepsilon} \qquad \text{(A-6)}$$

where $\mathbf{y}$ is the vector of observations, $\mathbf{X}$ is the matrix of levels of the variables, $\boldsymbol{\beta}$ is the vector of unknown coefficients, and $\boldsymbol{\varepsilon}$ is the vector of random errors. Afterwards, method of least squares, which minimizes the sum of squares of errors, is employed to find the estimators of the coefficients ($\hat{\boldsymbol{\beta}}$) through:

$$\hat{\boldsymbol{\beta}} = (\mathbf{X}'\mathbf{X})^{-1}\mathbf{X}'\mathbf{y} \qquad \text{(A-7)}$$

The fitted model will then be written as:

$$\hat{\mathbf{y}} = \mathbf{X}\hat{\boldsymbol{\beta}} \qquad \text{(A-8)}$$

Finally, response surfaces or contour plots are depicted to help visualize the relationship between the response and the variables and see the influence of the parameters. As you might notice, there is a close connection between RSM and linear regression analysis.

After the unknown coefficients (βs) were estimated by least squares method, the quadratic models for the mean fiber diameter (MFD) and standard deviation of fiber diameter (StdFD) in terms of coded variables are written as:

$$\begin{aligned}\mathrm{MFD} = {} & 282.031 + 34.953x_1 + 5.622x_2 - 2.113x_3 + 9.013x_4 \\ & - 11.613x_1^2 - 4.304x_2^2 - 15.500x_3^2 \\ & - 0.414x_4^2 + 12.517x_1x_2 + 4.020x_1x_3 - 0.162x_1x_4 + 20.643x_2x_3 + 0.741x_2x_4 + 0.877x_3x_4\end{aligned} \qquad \text{(A-9)}$$

$$\begin{aligned}\mathrm{StdFD} = {} & 36.1574 + 4.5788x_1 - 1.5536x_2 + 6.4012x_3 + 1.1531x_4 \\ & - 2.2937x_1^2 - 0.1115x_2^2 - 1.1891x_3^2 + 3.0980x_4^2 \\ & - 0.2088x_1x_2 + 1.0010x_1x_3 + 2.7978x_1x_4 + 0.1649x_2x_3 - 2.4876x_2x_4 + 1.5182x_3x_4\end{aligned} \qquad \text{(A-10)}$$

In the next step, a couple of very important hypothesis-testing procedures were carried out to measure the usefulness of the models presented here. First, the test for significance of the model was performed to determine whether there is a subset of variables which contributes significantly in representing the response variations. The appropriate hypotheses are:

$$H_0 : \beta_1 = \beta_2 = \cdots = \beta_k$$
$$H_1 : \beta_j \neq 0 \quad \text{for at least one } j \qquad \text{(A-11)}$$

The F statistics (the result of dividing the factor mean square by the error mean square) of this test along with the p-values (a measure of statistical significance, the smallest level of significance for which the null hypothesis is rejected) for both models are shown in Table 1.

Table A-1. Summary of the results from statistical analysis of the models

	F	p-value	R^2	R^2_{adj}	R^2_{pred}
MFD	106.02	0.000	95.74%	94.84%	93.48%
StdFD	42.05	0.000	89.92%	87.78%	84.83%

The p-values of the models are very small (almost zero), therefore it could be concluded that the null hypothesis is rejected in both cases suggesting that there are some significant terms in each model. There are also included in Table 1, the values of R^2, R^2_{adj}, and R^2_{pred}. R^2 is a measure for the amount of response variation which is explained by variables and will always increase when a new term is added to the model regardless of whether the inclusion of the additional term is statistically significant or not. R^2_{adj} is the adjusted form of R^2 for the number of terms in the model; therefore it will increase only if the new terms improve the model and decreases if unnecessary terms are added. R^2_{pred} implies how well the model predicts the response for new observations, whereas R^2 and R^2_{adj} indicate how well the model fits the experimental data. The R^2 values demonstrate that 95.74% of MFD and 89.92% of StdFD are explained by the variables. The R^2_{adj} values are 94.84% and 87.78% for MFD and StdFD respectively, which account for the number of terms in the models. Both R^2 and R^2_{adj} values indicate that the models fit the data very well. The slight difference between the values of R^2 and R^2_{adj} suggests that there might be some insignificant terms in the models. Since the R^2_{pred} values are so close to the values of R^2 and R^2_{adj}, models does not appear to be overfit and have very good predictive ability.

The second testing hypothesis is evaluation of individual coefficients, which would be useful for determination of variables in the models. The hypotheses for testing of the significance of any individual coefficient are:

$$H_0 : \beta_j = 0$$
$$H_1 : \beta_j \neq 0 \qquad \text{(A-12)}$$

The model might be more efficient with inclusion or perhaps exclusion of one or more variables. Therefore the value of each term in the model is evaluated using this test, and then eliminating the statistically insignificant terms, more efficient models could be obtained. The results of this test for the models of MFD and StdFD are summarized in Table 2 and Table 3 respectively. *T* statistic in these tables is a measure of the difference between an observed statistic and its hypothesized population value in units of standard error.

Table A-2. The test on individual coefficients for the model of mean fiber diameter (MFD)

Term (coded)	Coeff.	T	*p*-value
Constant	282.031	102.565	0.000
C	34.953	31.136	0.000
d	5.622	5.008	0.000
V	-2.113	-1.882	0.064
Q	9.013	8.028	0.000
C^2	-11.613	-5.973	0.000
d^2	-4.304	-2.214	0.030
V^2	-15.500	-7.972	0.000
Q^2	-0.414	-0.213	0.832
Cd	12.517	9.104	0.000
CV	4.020	2.924	0.005
CQ	-0.162	-0.118	0.906
dV	20.643	15.015	0.000
dQ	0.741	0.539	0.592
VQ	0.877	0.638	0.526

Table A-3. The test on individual coefficients for the model of standard deviation of fiber diameter (StdFD)

Term (coded)	Coef	T	*p*-value
Constant	36.1574	39.381	0.000
C	4.5788	12.216	0.000
D	-1.5536	-4.145	0.000
V	6.4012	17.078	0.000
Q	1.1531	3.076	0.003
C^2	-2.2937	-3.533	0.001
d^2	-0.1115	-0.172	0.864
V^2	-1.1891	-1.832	0.072
Q^2	3.0980	4.772	0.000
Cd	-0.2088	-0.455	0.651
CV	1.0010	2.180	0.033
CQ	2.7978	6.095	0.000
dV	0.1649	0.359	0.721
dQ	-2.4876	-5.419	0.000
VQ	1.5182	3.307	0.002

As depicted, the terms related to Q^2, CQ, dQ, and VQ in the model of MFD and related to d^2, Cd, and dV in the model of StdFD have very high *p*-values, therefore they do not contribute significantly in representing the variation of the corresponding response. Eliminating these terms will enhance the efficiency of the models. The new models are then given by recalculating the unknown coefficients in terms of coded variables in equations (A-13) and (A-14), and in terms of natural (uncoded) variables in equations (A-15) and (A-16).

$$MFD = 281.755 + 34.953x_1 + 5.622x_2 - 2.113x_3 + 9.013x_4 - 11.613x_1^2 - 4.304x_2^2 - 15.500x_3^2 + 12.517x_1x_2 + 4.020x_1x_3 + 20.643x_2x_3 \quad \text{(A-13)}$$

$$StdFD = 36.083 + 4.579x_1 - 1.554x_2 + 6.401x_3 + 1.153x_4 - 2.294x_1^2 - 1.189x_3^2 + 3.098x_4^2 + 1.001x_1x_3 + 2.798x_1x_4 - 2.488x_2x_4 + 1.518x_3x_4 \quad \text{(A-14)}$$

$$MFD = 10.3345 + 48.7288\,C - 22.7420\,d + 7.9713\,V + 90.1250\,Q - 2.9033\,C^2 - 0.1722\,d^2 - 0.6120\,V^2 + 1.2517\,Cd + 0.4020\,CV + 0.8257\,dV \quad \text{(A-15)}$$

$$StdFD = -1.8823 + 7.5590\,C + 1.1818\,d + 1.2709\,V - 300.3410\,Q - 0.5734\,C^2 - 0.0476\,V^2 + 309.7999\,Q^2 + 0.1001\,CV + 13.9892\,CQ - 4.9752\,dQ + 3.0364\,VQ \quad \text{(A-16)}$$

The results of the test for significance as well as R^2, R^2_{adj}, and R^2_{pred} for the new models are given in Table 4. It is obvious that the *p*-values for the new models are close to zero indicating the existence of some significant terms in each model. Comparing the results of this table with Table 1, the *F* statistic increased for the new models, indicating the improvement of the models after eliminating the insignificant terms. Despite the slight decrease in R^2, the values of R^2_{adj}, and R^2_{pred} increased substantially for the new models. As it was mentioned earlier in the paper, R^2 will always increase with the number of terms in the model. Therefore, the smaller R^2 values were expected for the new models, due to the fewer terms. However, this does not necessarily suggest that the pervious models were more efficient. Looking at the tables, R^2_{adj}, which provides a more useful tool for comparing the explanatory power of models with different number of terms, increased after eliminating the unnecessary variables. Hence, the new models have the ability to better explain the experimental data. Due to higher R^2_{pred}, the new models also have higher prediction ability. In other words, eliminating the insignificant terms results in simpler models which not only present the experimental data in superior form, but also are more powerful in predicting new conditions.

Table A-4. Summary of the results from statistical analysis of the models after eliminating the insignificant terms

	F	*p*-value	R^2	R^2_{adj}	R^2_{pred}
MFD	155.56	0.000	95.69%	95.08%	94.18%
StdFD	55.61	0.000	89.86%	88.25%	86.02%

The test for individual coefficients was performed again for the new models. The results of this test are summarized in Table 5 and Table 6. This time, as it was anticipated, no terms had higher *p*-value than expected, which need to be eliminated. Here is another advantage of removing unimportant terms. The values of T statistic increased for the terms already in the models implying that their effects on the response became stronger.

Table A-5. The test on individual coefficients for the model of mean fiber diameter (MFD) after eliminating the insignificant terms

Term (coded)	Coeff.	T	p-value
Constant	281.755	118.973	0.000
C	34.953	31.884	0.000
d	5.622	5.128	0.000
V	-2.113	-1.927	0.058
Q	9.013	8.221	0.000
C^2	-11.613	-6.116	0.000
d^2	-4.304	-2.267	0.026
V^2	-15.500	-8.163	0.000
Cd	12.517	9.323	0.000
CV	4.020	2.994	0.004
dV	20.643	15.375	0.000

Table A-6. The test on individual coefficients for the model of standard deviation of fiber diameter (StdFD) after eliminating the insignificant terms

Term (coded)	Coef	T	p-value
Constant	36.083	45.438	0.000
C	4.579	12.456	0.000
d	-1.554	-4.226	0.000
V	6.401	17.413	0.000
Q	1.153	3.137	0.003
C^2	-2.294	-3.602	0.001
V^2	-1.189	-1.868	0.066
Q^2	3.098	4.866	0.000
CV	1.001	2.223	0.029
CQ	2.798	6.214	0.000
dQ	-2.488	-5.525	0.000
VQ	1.518	3.372	0.001

After developing the relationship between parameters, the test data were used to investigate the prediction ability of the models. Root mean square errors (RMSE) between the calculated responses (C_i) and real responses (R_i) were determined using equation (17) for experimental data as well as test data for the sake of evaluation of both MFD and StdFD models.

$$\text{RMSE} = \sqrt{\frac{\sum_{i=1}^{n}(C_i - R_i)^2}{n}} \tag{A-17}$$

REFERENCES

[1] J. Doshi and D.H. Reneker, *J. Electrostatics*, 35, 151 (1995).

[2] H. Fong and D.H. Reneker, *Electrospinning and the formation of nanofibers*, in: D.R. Salem (Ed.), *Structure formation in polymeric fibers*, Hanser, Cincinnati (2001).

[3] D. Li and Y. Xia, *Adv. Mater.*, 16, 1151 (2004).

[4] R. Derch, A. Greiner and J.H. Wendorff, *Polymer nanofibers prepared by electrospinning*, in: J.A. Schwarz, C.I. Contescu and K. Putyera (Eds.), *Dekker encyclopedia of nanoscience and nanotechnology*, CRC, New York (2004).

[5] A.K. Haghi and M. Akbari, *Phys. Stat. Sol. A*, 204, 1830 (2007).

[6] P.W. Gibson, H.L. Schreuder-Gibson and D. Rivin, *AIChE J.*, 45, 190 (1999).

[7] M. Ziabari, V. Mottaghitalab and A.K. Haghi, *Korean J. Chem. Eng.*, 25, 923 (2008).

[8] Z.M. Huang, Y.Z. Zhang, M. Kotaki and S. Ramakrishna, *Compos. Sci. Technol.*, 63, 2223 (2003).

[9] M. Li, M.J. Mondrinos, M.R. Gandhi, F.K. Ko, A.S. Weiss and P.I. Lelkes, *Biomaterials*, 26, 5999 (2005).

[10] E.D. Boland, B.D. Coleman, C.P. Barnes, D.G. Simpson, G.E. Wnek and G.L. Bowlin, *Acta. Biomater.*, 1, 115 (2005).

[11] J. Lannutti, D. Reneker, T. Ma, D. Tomasko and D. Farson, *Mater. Sci. Eng. C*, 27, 504 (2007).

[12] J. Zeng, L. Yang, Q. Liang, X. Zhang, H. Guan, C. Xu, X. Chen and X. Jing, *J. Control. Release*, 105, 43 (2005).

[13] E.R. Kenawy, G.L. Bowlin, K. Mansfield, J. Layman, D.G. Simpson, E.H. Sanders and G.E. Wnek, *J. Control. Release*, 81, 57 (2002).

[14] M.S. Khil, D.I. Cha, H.-Y. Kim, I.-S. Kim and N. Bhattarai, *J. Biomed. Mater. Res. Part B: Appl. Biomater.*, 67, 675 (2003).

[15] B.M. Min, G. Lee, S.H. Kim, Y.S. Nam, T.S. Lee and W.H. Park, *Biomaterials*, 25, 1289 (2004).

[16] X.H. Qin and S.Y. Wang, *J. Appl. Polym. Sci.*, 102, 1285 (2006).

[17] H.S. Park and Y.O. Park, *Korean J. Chem. Eng.*, 22, 165 (2005).

[18] J.S. Kim and D.H. Reneker, *Poly. Eng. Sci.*, 39, 849 (1999).

[19] S.W. Lee, S.W. Choi, S.M. Jo, B.D. Chin, D.Y. Kim and K.Y. Lee, *J. Power Sources*, 163, 41 (2006).

[20] C. Kim, *J. Power Sources*, 142, 382 (2005).

[21] N.J. Pinto, A.T. Johnson, A.G. MacDiarmid, C.H. Mueller, N. Theofylaktos, D.C. Robinson and F.A. Miranda, *Appl. Phys. Lett.*, 83, 4244 (2003).

[22] D. Aussawasathien, J.-H. Dong and L. Dai, *Synthetic Met.*, 54, 37 (2005).

[23] [S.-Y. Jang, V. Seshadri, M.-S. Khil, A. Kumar, M. Marquez, P.T. Mather and G.A. Sotzing, *Adv. Mater.*, 17, 2177 (2005).

[24] S.-H. Tan, R. Inai, M. Kotaki and R. Ramakrishna, *Polymer*, 46, 6128 (2005).

[25] A, Ziabicki, *Fundamentals of fiber formation: The science of fiber spinning and drawing*, Wiley, New York (1976).

[26] A. Podgóski, A. Bałazy and L. Gradoń, *Chem. Eng. Sci.*, 61, 6804 (2006).

[27] B. Ding, M. Yamazaki and S. Shiratori, *Sens. Actuators B*, 106, 477 (2005).

[28] J.R. Kim, S.W. Choi, S.M. Jo, W.S. Lee and B.C. Kim, *Electrochim. Acta*, 50, 69 (2004).
[29] L. Moroni, R. Licht, J. de Boer, J.R. de Wijn and C.A. van Blitterswijk, *Biomaterials*, 27, 4911 (2006).

Chapter 3

SIMULATION ALGORITHM FOR GENERATION OF NANOWEBS

INTRODUCTION

Electrospinning is a novel and efficient method by which fibers with diameters in nanometer scale entitled as nanofibers, can be achieved. In electrospinning process, a strong electric field is applied on a droplet of polymer solution (or melt) held by its surface tension at the tip of a syringe's needle (or a capillary tube). As a result, the pendent drop will become highly electrified and the induced charges are distributed over its surface. Increasing the intensity of electric field, the surface of the liquid drop will be distorted to a conical shape known as the Taylor cone [1]. Once the electric field strength exceeds a threshold value, the repulsive electric force dominates the surface tension of the liquid and a stable jet emerges from the cone tip. The charged jet is then accelerated toward the target and rapidly thins and dries as a result of elongation and solvent evaporation. As the jet diameter decreases, the surface charge density increases and the resulting high repulsive forces split the jet to smaller jets. This phenomenon may take place several times leading to many small jets. Ultimately, solidification is carried out and fibers are deposited on the surface of the collector as a randomly oriented nonwoven mat [2]-[5]. Figure 1 shows a schematic illustration of electrospinning setup.

According to various outstanding properties such as very small fiber diameters, large surface area per mass ratio [3], high porosity along with small pore sizes [7], [8], flexibility, and superior mechanical properties [9], electrospun nanofiber mats have found numerous applications in diverse areas. For example in biomedical field nanofibers plays a substantial role in tissue engineering [10]-[12], drug delivery [13], [14], and wound dressing [15], [16]. Moreover, the use of nanofibers in protective clothing [7], filtration technology [17], [18] and reinforcement of composite materials [9], [19] is extremely significant for developing of specific products by manipulation of materials in nanoscales. In the mean time, those applications related to micro-electronics like battery [20], supercapacitors [21], transistors [22], sensors [23], and display devices [24]

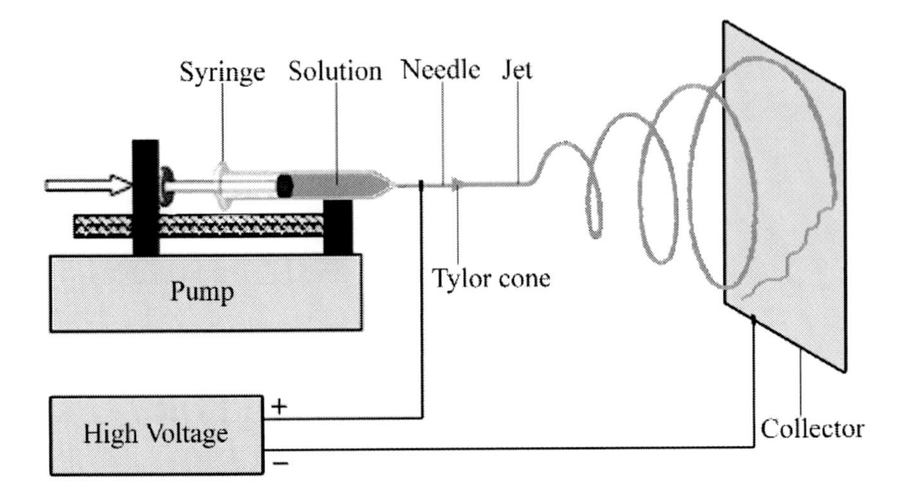

Figure 1. A typical image of Electrospinning process [6].

The physical characteristics of electrospun nanofibers such as fiber diameter depend on various parameters which are mainly divided into three categories: solution properties (solution viscosity, solution concentration, polymer molecular weight, and surface tension), processing conditions (applied voltage, volume flow rate, spinning distance, and needle diameter), and ambient conditions (temperature, humidity, and atmosphere pressure) [25]. Numerous applications require nanofibers with desired properties suggesting the importance of the process control. This end may not be achieved unless having a comprehensive outlook of the process and quantitative study of the effects of governing parameters. In addition, qualitative description of the experimental observations are not adequate to derive general conclusions and either the equations governing behavior of the system must be found or appropriate empirical models need be presented. In Ziabicki's words, "in the language of science 'to explain' means to put forward a quantitative model which is consistent with all the known data and capable of predicting new fact" [26].

Employing a model to express the influence of electrospinning parameters will help us obtain a simple and systematic way for presenting the effects of variables thereby enabling the control of the process. Furthermore, it allows us to predict the results under new combination of parameters. Hence, without conducting any experiments, one can easily estimate features of the product under unknown conditions. An empirical model therefore tells us to what extent the output of a system will change if one or more parameters increased or decreased. This is very helpful and leads to a complete understanding of the process and the effects of parameters.

Despite the surge in attention to different aspects of electrospinning process, a few investigations have addressed the quantitative study of the effects of the parameters. Changing the behavior of materials in nano-scale, presence of electric field, branching of the jet, random orientation of fibers, etc. made the analysis of the process extremely complex and difficult that to date there has been no reliable theory capable of describing the phenomenon. Furthermore, the development of an empirical model has also been impeded due to the lack of systematic and characterized experimentations with appropriate designs. Adding to the difficulty is the number of parameters involving in the electrospinning process and the interactions between them which made it almost impossible to investigate the simultaneous effects of all variables.

Affecting the characteristics of the final product such as physical, mechanical and electrical properties, fiber diameter is one of the most important structural features in electrospun nanofiber mats. Podgorski et al. [27] indicated that filters composed of fibers with smaller diameters have higher filtration efficiencies. This was also proved by the work presented by Qin et al. [17]. Ding et al. [28] also reported that sensitivity of sensors increase with decreasing the mean fiber diameter due to the higher surface area. It was also shown that in polymer batteries consisting of electrospun polyvinylidene fluoride (PVdF) fibrous electrolyte, lower mean fiber diameter results in a higher electrolyte uptake thereby increasing ionic conductivity of the mat [29]. Furthermore, Moroni et al. [30] found that fiber diameters of electrospun polyethylene oxide terephthalate/polybutylene terephthalate (PEOT/PBT) scaffolds influencing on cell seeding, attachment, and proliferation. They also studied the release of dye incorporated in electrospun scaffolds and observed that with increasing fiber diameter, the cumulative release of the dye (methylene blue) decreased. Carbonization and activation conditions as well as the structure and properties of the ultimate carbon fibers are also affected by the diameters of the precursor polyacrylonitril (PAN) nanofibers [31]. Consequently, precise control of the electrospun fiber diameter is very crucial.

A few techniques such as orthogonal experimental design [32] and using power law relationships [31] have been reported in the literature for quantitative study of electrospun nanofiber. However researchers mostly paid attention to response surface methodology (RSM) technique due to its simplicity and its ability to take into account the interactions between the parameters. Sukigara et al. [33] employed RSM to model mean fiber diameter of electrospun regenerated Bombyx mori silk with electric field and concentration at two spinning distances. They applied a full factorial experimental design at three levels of each parameter leading to nine treatments of factors and used a quadratic polynomial to establish a relationship between mean fiber diameter and the variables. Increasing the concentration at constant electric field resulted in an increase in mean fiber diameter. Different impacts for the electric field were observed depending on solution concentration. Since the effects of solution concentration and electric field strength on mean fiber diameter changed at different spinning distances, they suggested that some interactions and coupling effects are present between the parameters.

Gu et al. [34], [35] also exploited the RSM for quantitative study of polyacrylonitril (PAN) and poly D,L-lactide (PDLA) respectively. The only difference observed in the procedure was the use of four levels of concentration in the former case. They included the standard deviation of fiber diameter in their investigations by which they were able to provide additional information regarding the morphology of electrospun nanofibers and its variations at different conditions. Furthermore, they analyzed the significance of factors in the models in order to understand the level of influence of each parameter. In the case of PAN, voltage as well as its interaction with concentration had no considerable effects on both mean and standard deviation of fiber diameter. Hence, they eliminated the terms corresponding to these factors thereby obtained simplified quadratic models according to which mean and standard deviation of fiber diameter increased with polymer concentration. On the contrary, both voltage and its interaction with concentration were found to be significant in the case of PDLA. However, the effect of polymer concentration was more pronounced. Increasing voltage at constant concentration favored thinner fiber formation which gained momentum with increasing concentration. Fibers with more uniform diameters (less standard deviation) were obtained at higher applied voltage or concentration.

In the most recent investigation in this field, Yördem et al. [36] utilized RSM to correlate mean and coefficient of variation (CV) of electrospun PAN nanofibers to solution concentration and applied voltage at three different spinning distances. They employed a face-centered central composite design (FCCD) along with a full factorial design at two levels resulting in 13 treatments at each spinning distance. A cubic polynomial was then used to fit the data in each case. As previous studies, fiber diameter was very sensitive to changes in solution concentration. Voltage effect was more significant at higher concentrations demonstrating the interaction between parameters.

According to previous studies, there are some interactions between electrospinning parameters. However, they only investigated the simultaneous effects of two variables; therefore they were unable to thoroughly capture the interactions which exist between the parameters. For instance, Sukigara et al. [33] and Yördem et al. [36] both agreed that spinning distance has a significant influence on fiber diameter and that this effect varies when solution concentration and/or applied voltage altered. Although, their study promotes our knowledge about quantitative analysis of electrospinning process, still suffer from lack of comprehensiveness. In addition, in every research where modeling of a process is targeted, the obtained models need to be evaluated with a set of test data which were not used in establishing the relationships. Otherwise, the effectiveness of the models will not be guaranteed and there will always be an uncertainty in the prediction of the models in new conditions. Therefore it could be claimed that, the presented models in previous studies were not evaluated with a series of test data. Therefore, their models may not generalize well to new data and their prediction ability is obscure.

In this contribution for the first time, the simultaneous effects of four electrospinning parameters (solution concentration, spinning distance, applied voltage, and volume flow rate) on mean and standard deviation of electrospun polyvinyl alcohol (PVA) fiber diameter were systematically investigated. PVA, the largest volume synthetic water-soluble polymer produced in the world, is commercially manufactured by the hydrolysis of polyvinyl acetate. The excellent chemical resistance and physical properties of PVA along with non-toxicity and biodegradability have led to its broad industrial applications such as textile sizing, adhesive, paper coating, fibers, and polymerization stabilizers [37], [38]. Several patents reported process for production of ultrahigh tensile strength PVA fibers comparable to Kevlar® [39]-[41]. PVA has found many applications in biomedical uses as well, due to its biocompatibility [42]. For instance, PVA hydrogels were used in regenerating articular cartilages [43], [44], artificial pancreas [45], and drug delivery systems [46], [47]. More recently, PVA nanofibers were electrospun and used as a protein delivery system [48], retardation of enzyme release [48] and wound dressing [49]. The objective of this paper is to use RSM to establish quantitative relationships between electrospinning parameters and mean and standard deviation of fiber diameter as well as to evaluate the effectiveness of the empirical models with a set of test data.

EXPERIMENTAL

Solution Preparation and Electrospinning

PVA with molecular weight of 72000 *g/mol* and degree of hydrolysis of >98% was obtained from Merck and used as received. Distilled water as solvent was added to a predetermined amount of PVA powder to obtain 20 *ml* of solution with desired concentration. The solution was prepared at 80°C and gently stirred for 30 min to expedite the dissolution. After the PVA had completely dissolved, the solution was transferred to a 5 *ml* syringe and became ready for spinning of nanofibers. The experiments were carried out on a horizontal electrospinning setup shown schematically in Figure 1. The syringe containing PVA solution was placed on a syringe pump (New Era NE-100) used to dispense the solution at a controlled rate. A high voltage DC power supply (Gamma High Voltage ES-30) was used to generate the electric field needed for electrospinning. The positive electrode of the high voltage supply was attached to the syringe needle via an alligator clip and the grounding electrode was connected to a flat collector wrapped with aluminum foil where electrospun nanofibers were accumulated to form a nonwoven mat. The electrospinning was carried out at room temperature. Subsequently, the aluminum foil was removed from the collector. A small piece of mat was placed on the sample holder and gold sputter-coated (Bal-Tec). Thereafter, the micrograph of electrospun PVA fibers was obtained using scanning electron microscope (SEM, Phillips XL-30) under magnification of 10000X. Quite recently, the authors established a couple of image analysis based techniques entitled as *direct tracking* [50], [51] and *new distance transform* [52], [53] for measuring electrospun nanofiber diameter. In this study, fiber diameter distribution for each specimen was determined from the SEM micrograph by new distance transform method due to its effectiveness. SEM micrographs of typical PVA electrospun nanofiber mats are shown in Figure 4 in Appendix.

Choice of Parameters and Range

Variables which potentially can alter the electrospinning process are large. Hence, investigating all of them in the framework of one single research would almost be impossible. However, some of these parameters can be held constant during experimentation. For instance, performing the experiments in a controlled environmental condition, which is concerned in this study, the ambient parameters (i.e. temperature, air pressure, and humidity) are kept unchanged. Solution viscosity is affected by polymer molecular weight, solution concentration, and temperature. For a particular polymer (constant molecular weight) at a fixed temperature, solution concentration would be the only factor influencing the viscosity. In this circumstance, the effect of viscosity could be determined by the solution concentration. Therefore, there would be no need for viscosity to be considered as a separate parameter.

In this regard, solution concentration (C), spinning distance (d), applied voltage (V), and volume flow rate (Q) were selected to be the most influential parameters. The next step is to choose the ranges over which these factors are varied. Process knowledge, which is a combination of practical experience and theoretical understanding, is required to fulfill this step. The aim is here to find an appropriate range for each parameter where dry, bead-free, stable, and continuous fibers without breaking up to droplets are obtained. This goal could be achieved by conducting a set of preliminary experiments while having the previous works in mind along with utilizing the reported relationships.

The relationship between intrinsic viscosity ($[\eta]$) and molecular weight (M) is given by the well-known Mark-Houwink-Sakurada equation as follows:

$$[\eta] = KM^{a} \quad (1)$$

where K and a are constants for a particular polymer-solvent pair at a given temperature [54]. For the PVA with molecular weight in the range of 69000 g/mol$<$M$<$690000 g/mol in water at room temperature, K=6.51 and a=0.628 were found by Tacx et al. [55]. Using these constants in the equation, the intrinsic viscosity for PVA in this study (molecular weight of 72000 g/mol) was calculated to be $[\eta]=0.73$.

Polymer chain entanglements in a solution can be expressed in terms of Berry number (B), which is a dimensionless parameter and defined as the product of intrinsic viscosity and polymer concentration ($B=[\eta]C$) [56]. For each molecular weight, there is a lower critical concentration at which the polymer solution cannot be electrospun. Koski et al. [57] observed that $B>5$ is required to form stabilized fibrous structures in electrospinning of PVA. On the other hand, they reported the formation of flat fibers at $B>9$. Therefore, the appropriate range in this case could be found within $5<B<9$ domain which is equivalent to $6.8\%<C<12.3\%$ in terms of concentration of PVA. Furthermore, Koski et al. [57] observed that beaded fibers were electrospun at low solution concentration. Hence, it was thought that the domain $8\%\leq C\leq 12\%$ would warrant the formation of stabilized bead-free fibers with circular cross-sections. This domain was later justified by performing some preliminary experiments.

As for determining the appropriate range of applied voltage, referring to previous works, it was observed that the changes of voltage lay between 5 kV to 25 kV depending on experimental conditions; voltages above 25 kV were rarely used. Afterwards, a series of experiments were carried out to obtain the desired voltage domain. At $V<10$ kV, the voltage was too low to spin fibers and 10 $kV\leq V<15$ kV resulted in formation of fibers and droplets; in addition, electrospinning was impeded at high concentrations. In this regard, 15 $kV\leq V\leq 25$ kV was selected to be the desired domain for applied voltage.

The use of 5 cm – 20 cm for spinning distance was reported in the literature. Short distances are suitable for highly evaporative solvents whereas it results in wet coagulated fibers for nonvolatile solvents due to insufficient evaporation time. Since water was used as solvent for PVA in this study, short spinning distances were not expected to be favorable for dry fiber formation. Afterwards, this was proved by experimental observations and 10 $cm\leq d\leq 20$ cm was considered as the effective range for spinning distance.

Few researchers have addressed the effect of volume flow rate. Therefore in this case, the attention was focused on experimental observations. At $Q<0.2$ ml/h, in most cases especially at high polymer concentrations, the fiber formation was hindered due to insufficient supply of solution to the tip of the syringe needle. Whereas, excessive feed of solution at $Q>0.4$ ml/h incurred formation of droplets along with fibers. As a result, 0.2 $ml/h\leq Q\leq 0.4$ ml/h was chosen as the favorable range of flow rate in this study.

Experimental Design

The aim of experimental design is to provide reasonable and scientific answers to such questions. In other words, experimental design comprise sequential steps to ensure efficient data gathering process and leading to valid statistical inferences [58], [59].

Consider a process in which several factors affect a response of the system. In this case, a conventional strategy of experimentation, which is extensively used in practice, is the *one-factor-at-a-time* approach. The major disadvantage of this approach is its failure to consider any possible interaction between the factors, say the failure of one factor to produce the same effect on the response at different levels of another factor. For instance, suppose that two factors A and B affect a response. At one level of A, increasing B causes the response to increase, while at the other level of A, the effect of B totally reverses and the response decreases with increasing B. As interactions exist between electrospinning parameters, this approach may not be an appropriate choice for the case of the present work. The correct strategy to deal with several factors is to use a full factorial design. In this method, factors are all varied together; therefore all possible combinations of the levels of the factors are investigated. This approach is very efficient, makes the most use of the experimental data and takes into account the interactions between factors [58], [59].

It is trivial that in order to draw a line at least two points and for a quadratic curve at least three points are required. Hence, three levels were selected for each parameter in this study so that it would be possible to use quadratic models. These levels were chosen equally spaced. A full factorial experimental design with four factors (solution concentration, spinning distance, applied voltage, and flow rate) each at three levels (3^4 design) were employed resulting in 81 treatment combinations. This design is shown in Figure A-32.

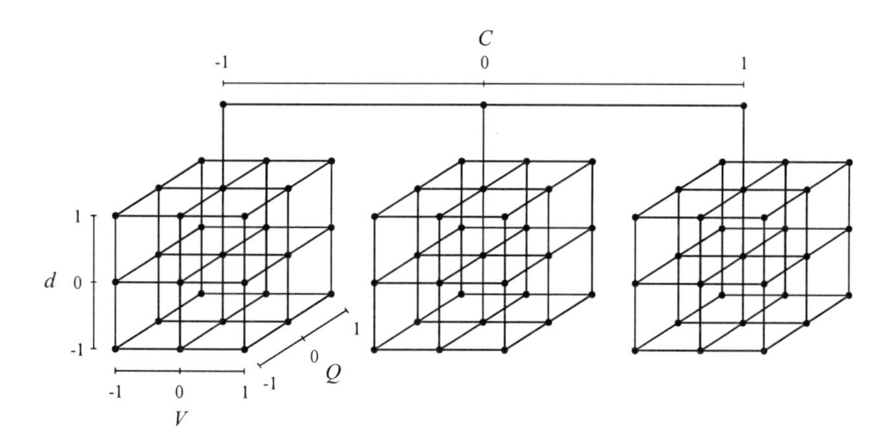

Figure 2. 3^4 full factorial experimental design used in this study.

-1, 0, and 1 are coded variables corresponding to low, intermediate and high levels of each factor respectively. The coded variables (x_j) were calculated using Equation (2) from natural variables (ξ_i). The indices 1 to 4 represent solution concentration, spinning distance, applied voltage, and flow rate respectively. In addition to experimental data, 15 treatments inside the design space were selected as test data and used for evaluation of the models. The natural and coded variables for experimental data (numbers 1-81) as well as test data (numbers 82-96) are listed in Table 8 in Appendix.

$$x_j = \frac{\xi_j - [\xi_{hj} + \xi_{lj}]/2}{[\xi_{hj} - \xi_{lj}]/2} \tag{2}$$

Response Surface Methodology

The mechanism of some scientific phenomena has been well understood and models depicting the physical behavior of the system have been drawn in the form of mathematical relationships. However, there are numerous processes at the moment which have not been sufficiently been understood to permit the theoretical approach. Response surface methodology (RSM) is a combination of mathematical and statistical techniques useful for empirical modeling and analysis of such systems. The application of RSM is in situations where several input variables are potentially influence some performance measure or quality characteristic of the process – often called responses. The relationship between the response (y) and k input variables ($\xi_1,\xi_2,...,\xi_k$) could be expressed in terms of mathematical notations as follows:

$$y = f(\xi_1, \xi_2, ..., \xi_k) \tag{3}$$

where the true response function f is unknown. It is often convenient to use coded variables ($x_1, x_2, .., x_k$) instead of natural (input) variables. The response function will then be:

$$y = f(x_1, x_2, ..., x_k) \tag{4}$$

Since the form of true response function f is unknown, it must be approximated. Therefore, the successful use of RSM is critically dependent upon the choice of appropriate function to approximate f. Low-order polynomials are widely used as approximating functions. First order (linear) models are unable to capture the interaction between parameters which is a form of curvature in the true response function. Second order (quadratic) models will be likely to perform well in these circumstances. In general, the quadratic model is in the form of:

$$y = \beta_0 + \sum_{j=1}^{k} \beta_j x_j + \sum_{j=1}^{k} \beta_{jj} x_j^2 + \sum_{i<j} \sum_{j=2}^{k} \beta_{ij} x_i x_j + \varepsilon \tag{5}$$

where ε is the error term in the model. The use of polynomials of higher order is also possible but infrequent. The βs are a set of unknown coefficients needed to be estimated. In order to do that, the first step is to make some observations on the system being studied. The model in Equation (5) may now be written in matrix notations as:

$$\mathbf{y} = \mathbf{X}\boldsymbol{\beta} + \boldsymbol{\varepsilon} \tag{6}$$

where **y** is the vector of observations, **X** is the matrix of levels of the variables, **β** is the vector of unknown coefficients, and **ε** is the vector of random errors. Afterwards, method of least squares, which minimizes the sum of squares of errors, is employed to find the estimators of the coefficients ($\hat{\boldsymbol{\beta}}$) through:

$$\hat{\boldsymbol{\beta}} = (\mathbf{X}'\mathbf{X})^{-1}\mathbf{X}'\mathbf{y} \tag{7}$$

The fitted model will then be written as:

$$\hat{\mathbf{y}} = \mathbf{X}\hat{\boldsymbol{\beta}} \tag{8}$$

Finally, response surfaces or contour plots are depicted to help visualize the relationship between the response and the variables and see the influence of the parameters [60], [61]. As you might notice, there is a close connection between RSM and linear regression analysis [62].

In this study RSM was employed to establish empirical relationships between four electrospinning parameters (solution concentration, spinning distance, applied voltage, and flow rate) and two responses (mean fiber diameter and standard deviation of fiber diameter). Coded variables were used to build the models. The choice of three levels for each factor in experimental design allowed us to take the advantage of quadratic models. Afterwards, the significance of terms in each model was investigated by testing hypotheses on individual coefficients and simpler yet more efficient models were obtained by eliminating statistically unimportant terms. Finally, the validity of the models was evaluated using the 15 test data. The analyses were carried out using statistical software Minitab 15.

RESULTS AND DISCUSSION

After the unknown coefficients (*β*s) were estimated by least squares method, the quadratic models for the mean fiber diameter (MFD) and standard deviation of fiber diameter (StdFD) in terms of coded variables are written as:

$$\begin{aligned} \text{MFD} = {} & 282.031 + 34.953x_1 + 5.622x_2 - 2.113x_3 + 9.013x_4 \\ & - 11.613x_1^2 - 4.304x_2^2 - 15.500x_3^2 \\ & - 0.414x_4^2 + 12.517x_1x_2 + 4.020x_1x_3 - 0.162x_1x_4 + 20.643x_2x_3 + 0.741x_2x_4 + 0.877x_3x_4 \end{aligned} \tag{9}$$

$$\begin{aligned} \text{StdFD} = {} & 36.1574 + 4.5788x_1 - 1.5536x_2 + 6.4012x_3 + 1.1531x_4 \\ & - 2.2937x_1^2 - 0.1115x_2^2 - 1.1891x_3^2 + 3.0980x_4^2 \\ & - 0.2088x_1x_2 + 1.0010x_1x_3 + 2.7978x_1x_4 + 0.1649x_2x_3 - 2.4876x_2x_4 + 1.5182x_3x_4 \end{aligned} \tag{10}$$

In the next step, a couple of very important hypothesis-testing procedures were carried out to measure the usefulness of the models presented here. First, the test for significance of the model was performed to determine whether there is a subset of variables which contributes significantly in representing the response variations. The appropriate hypotheses are:

$$H_0 : \beta_1 = \beta_2 = \cdots = \beta_k$$
$$H_1 : \beta_j \neq 0 \quad \text{for at least one } j \tag{11}$$

The F statistics (the result of dividing the factor mean square by the error mean square) of this test along with the p-values (a measure of statistical significance, the smallest level of significance for which the null hypothesis is rejected) for both models are shown in Table 1.

Table1. Summary of the results from statistical analysis of the models

	F	p-value	R^2	R^2_{adj}	R^2_{pred}
MFD	106.02	0.000	95.74%	94.84%	93.48%
StdFD	42.05	0.000	89.92%	87.78%	84.83%

The p-values of the models are very small (almost zero), therefore it could be concluded that the null hypothesis is rejected in both cases suggesting that there are some significant terms in each model. There are also included in Table 1, the values of R 2, R 2adj, and R 2pred. R 2 is a measure for the amount of response variation which is explained by variables and will always increase when a new term is added to the model regardless of whether the inclusion of the additional term is statistically significant or not. R 2adj is the adjusted form of R 2 for the number of terms in the model; therefore it will increase only if the new terms improve the model and decreases if unnecessary terms are added. R 2pred implies how well the model predicts the response for new observations, whereas R 2 and R 2adj indicate how well the model fits the experimental data. The R 2 values demonstrate that 95.74% of MFD and 89.92% of StdFD are explained by the variables. The R 2adj values are 94.84% and 87.78% for MFD and StdFD respectively, which account for the number of terms in the models. Both R 2 and R 2adj values indicate that the models fit the data very well. The slight difference between the values of R 2 and R 2adj suggests that there might be some insignificant terms in the models. Since the R 2pred values are so close to the values of R 2 and R 2adj, models does not appear to be overfit and have very good predictive ability.

The second testing hypothesis is evaluation of individual coefficients, which would be useful for determination of variables in the models. The hypotheses for testing of the significance of any individual coefficient are:

$$H_0 : \beta_j = 0$$
$$H_1 : \beta_j \neq 0 \tag{12}$$

The model might be more efficient with inclusion or perhaps exclusion of one or more variables. Therefore the value of each term in the model is evaluated using this test, and then eliminating the statistically insignificant terms, more efficient models could be obtained. The results of this test for the models of MFD and StdFD are summarized in Table 2 and Table 3 respectively. T statistic in these tables is a measure of the difference between an observed statistic and its hypothesized population value in units of standard error.

Table 2. The test on individual coefficients for the model of mean fiber diameter (MFD)

Term (coded)	Coef	T	p-value
Constant	282.031	102.565	0.000
C	34.953	31.136	0.000
d	5.622	5.008	0.000
V	-2.113	-1.882	0.064
Q	9.013	8.028	0.000
C^2	-11.613	-5.973	0.000
d^2	-4.304	-2.214	0.030
V^2	-15.500	-7.972	0.000
Q^2	-0.414	-0.213	0.832
Cd	12.517	9.104	0.000
CV	4.020	2.924	0.005
CQ	-0.162	-0.118	0.906
dV	20.643	15.015	0.000
dQ	0.741	0.539	0.592
VQ	0.877	0.638	0.526

Table 3. The test on individual coefficients for the model of standard deviation of fiber diameter (StdFD)

Term (coded)	Coef	T	p-value
Constant	36.1574	39.381	0.000
C	4.5788	12.216	0.000
D	-1.5536	-4.145	0.000
V	6.4012	17.078	0.000
Q	1.1531	3.076	0.003
C^2	-2.2937	-3.533	0.001
d^2	-0.1115	-0.172	0.864
V^2	-1.1891	-1.832	0.072
Q^2	3.0980	4.772	0.000
Cd	-0.2088	-0.455	0.651
CV	1.0010	2.180	0.033
CQ	2.7978	6.095	0.000
dV	0.1649	0.359	0.721
dQ	-2.4876	-5.419	0.000
VQ	1.5182	3.307	0.002

As depicted, the terms related to Q^2, CQ, dQ, and VQ in the model of MFD and related to d^2, Cd, and dV in the model of StdFD have very high p-values, therefore they do not contribute significantly in representing the variation of the corresponding response. Eliminating these terms will enhance the efficiency of the models. The new models are then given by recalculating the unknown coefficients in terms of coded variables in equations (13) and (14), and in terms of natural (uncoded) variables in equations (15), (16).

$$MFD = 281.755 + 34.953x_1 + 5.622x_2 - 2.113x_3 + 9.013x_4 - 11.613x_1^2 - 4.304x_2^2 - 15.500x_3^2 + 12.517x_1x_2 + 4.020x_1x_3 + 20.643x_2x_3 \quad (13)$$

$$StdFD = 36.083 + 4.579x_1 - 1.554x_2 + 6.401x_3 + 1.153x_4 - 2.294x_1^2 - 1.189x_3^2 + 3.098x_4^2 + 1.001x_1x_3 + 2.798x_1x_4 - 2.488x_2x_4 + 1.518x_3x_4 \quad (14)$$

$$MFD = 10.3345 + 48.7288C - 22.7420d + 7.9713V + 90.1250Q - 2.9033C^2 - 0.1722d^2 - 0.6120V^2 + 1.2517Cd + 0.4020CV + 0.8257dV \quad (15)$$

$$StdFD = -1.8823 + 7.5590C + 1.1818d + 1.2709V - 300.3410Q - 0.5734C^2 - 0.0476V^2 + 309.7999Q^2 + 0.1001CV + 13.9892CQ - 4.9752dQ + 3.0364VQ \quad (16)$$

The results of the test for significance as well as R^2, R^2_{adj}, and R^2_{pred} for the new models are given in Table 4. It is obvious that the *p*-values for the new models are close to zero indicating the existence of some significant terms in each model. Comparing the results of this table with Table 1, the *F* statistic increased for the new models, indicating the improvement of the models after eliminating the insignificant terms. Despite the slight decrease in R^2, the values of R^2_{adj}, and R^2_{pred} increased substantially for the new models. As it was mentioned earlier in the paper, R^2 will always increase with the number of terms in the model. Therefore, the smaller R^2 values were expected for the new models, due to the fewer terms. However, this does not necessarily suggest that the pervious models were more efficient. Looking at the tables, R^2_{adj}, which provides a more useful tool for comparing the explanatory power of models with different number of terms, increased after eliminating the unnecessary variables. Hence, the new models have the ability to better explain the experimental data. Due to higher R^2_{pred}, the new models also have higher prediction ability. In other words, eliminating the insignificant terms results in simpler models which not only present the experimental data in superior form, but also are more powerful in predicting new conditions. In the study conducted by Yördem et al. [36], despite high reported R^2 values, the presented models seem to be inefficient and uncertain. Some terms in the models had very high *p*-values. For instance, in modeling the mean fiber diameter, *p*-value as high as 0.975 was calculated for cubic concentration term at spinning distance of 16 *cm*, where half of the terms had *p*-values more than 0.8. This results in low R^2_{pred} values which were not reported in their study and after calculating by us, they were found to be almost zero in many cases suggesting the poor prediction ability of their models.

Table 4. Summary of the results from statistical analysis of the models after eliminating the insignificant terms

	F	*p*-value	R^2	R^2_{adj}	R^2_{pred}
MFD	155.56	0.000	95.69%	95.08%	94.18%
StdFD	55.61	0.000	89.86%	88.25%	86.02%

The test for individual coefficients was performed again for the new models. The results of this test are summarized in Table 5 and Table 6. This time, as it was anticipated, no terms had higher *p*-value than expected, which need to be eliminated. Here is another advantage of removing unimportant terms. The values of *T* statistic increased for the terms already in the models implying that their effects on the response became stronger.

Table 5. The test on individual coefficients for the model of mean fiber diameter (MFD) after eliminating the insignificant terms

Term (coded)	Coef	T	*p*-value
Constant	281.755	118.973	0.000
C	34.953	31.884	0.000
d	5.622	5.128	0.000
V	-2.113	-1.927	0.058
Q	9.013	8.221	0.000
C^2	-11.613	-6.116	0.000
d^2	-4.304	-2.267	0.026
V^2	-15.500	-8.163	0.000
Cd	12.517	9.323	0.000
CV	4.020	2.994	0.004
dV	20.643	15.375	0.000

Table 6. The test on individual coefficients for the model of standard deviation of fiber diameter (StdFD) after eliminating the insignificant terms

Term (coded)	Coef	T	*p*-value
Constant	36.083	45.438	0.000
C	4.579	12.456	0.000
d	-1.554	-4.226	0.000
V	6.401	17.413	0.000
Q	1.153	3.137	0.003
C^2	-2.294	-3.602	0.001
V^2	-1.189	-1.868	0.066
Q^2	3.098	4.866	0.000
CV	1.001	2.223	0.029
CQ	2.798	6.214	0.000
dQ	-2.488	-5.525	0.000
VQ	1.518	3.372	0.001

After developing the relationship between parameters, the test data were used to investigate the prediction ability of the models. Root mean square errors (RMSE) between the calculated responses (C_i) and real responses (R_i) were determined using equation (17) for experimental data as well as test data for the sake of evaluation of both MFD and StdFD models and the results are listed in Table 7. The models present acceptable RMSE values for test data indicating the ability of the models to generalize well the experimental data to predicting new conditions. Although the values of RMSE for the test data are slightly higher than experimental data, these small discrepancies were expected since it is almost impossible for an empirical model to express the test data as well as experimental data and higher errors are often obtained when new data are presented to the models. Hence, the results imply the acceptable prediction ability of the models.

$$\text{RMSE} = \sqrt{\frac{\sum_{i=1}^{n}(C_i - R_i)^2}{n}} \tag{17}$$

Table 7. RMSE values of the models for the experimental and test data

	Experimental data	Test data
MFD	7.489	10.647
StdFD	2.493	2.890

RESPONSE SURFACES FOR MEAN FIBER DIAMETER

Solution Concentration

A monotonic increase in MFD with concentration was observed in this study as shown in Figure (a), (b), and (c) which concurs with the previous observations [25], [31], [63]-[65]. The concentration effect was more pronounced at further spinning distances (Figure (a)). This could be attributed to the twofold effect of distance. At low concentrations, higher solvent content in the solution and longer distance provides more time not only to stretch the jet in the electric field but also to evaporate the solvent, thereby encouraging thinner fiber formation. At higher concentrations, however, there are extensive polymer chain entanglements resulting in higher viscoelastic forces which tend to resist against the electrostatic stretching force. On the other hand, increasing the spinning distance will reduce the electric field strength ($E=V/d$) causing the electrostatic force to decrease. As a result, increasing MFD with concentration gains more momentum at longer spinning distances. Higher applied voltages also accelerate the concentration impact on MFD (Figure (b)) which may be ascribed to the two fold effect of voltage. At higher voltages, where the electric field is strong and dominant factor, increasing polymer concentration tends to encourage the effect of voltage on mass flow rate of polymer. Hence, more solution could be removed from the tip of the needle resulting in further increase in MFD. No combined effect between solution concentration and volume flow rate was observed as depicted in Figure (c). Therefore, concentration had interactions with spinning distance and applied voltage which had been suggested by the existence of terms *Cd* and *CV* in the model of MFD. Recall that the term *CQ* was statistically insignificant and therefore had been removed from the model of MFD.

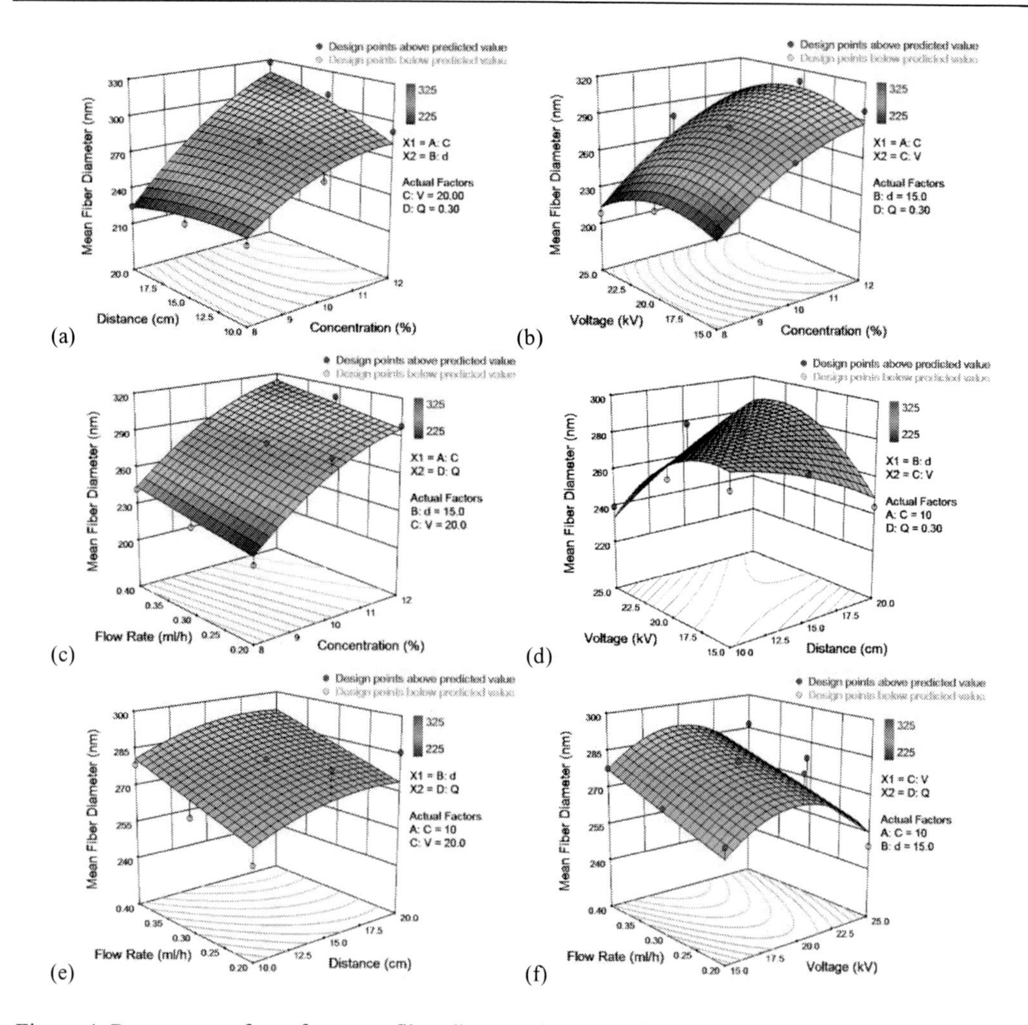

Figure 4. Response surfaces for mean fiber diameter in terms of: (a) solution concentration and spinning distance, (b) solution concentration and applied voltage, (c) solution concentration and flow rate, (d) spinning distance and applied voltage, (e) spinning distance and flow rate, (f) applied voltage and flow rate.

Spinning Distance

The impact of spinning distance on MFD is illustrated in Figure (a), (d), and (e). As it is depicted in these figures, the effect of spinning distance is not always the same. Spinning distance has a twofold effect on electrospun fiber diameter. Varying the distance has a direct influence on the jet flight time as well as electric field strength. Longer spinning distance will provide more time for the jet to stretch in the electric field before it is deposited on the collector. Furthermore, solvents will have more time to evaporate. Hence, the fiber diameter will be prone to decrease. On the other hand, increasing the spinning distance, the electric field strength will decrease ($E=V/d$) resulting in less acceleration hence stretching of the jet which leads to thicker fiber formation. The balance between these two effects will determine the final fiber diameter. Increase in fiber diameter diameter [64], [67], [68] as well as decrease in fiber diameter [31] with increasing spinning distance was reported in the literature. There were also some cases in which spinning distance did not have a significant influence on fiber diameter [63], [69]-[71]. As mentioned before, there will be more chain entanglements at higher concentrations resulting in an increase in viscoelastic force. Furthermore, the longer the distance, the lower is the electric field strength. Hence, the electrostatic stretching force, which has now become weaker, will be dominated easier by the viscoelastic force. As a result, the effect of spinning distance on fiber diameter is more highlighted, rendering higher MFD (Figure (a)). The effect of spinning distance will also alter at different applied voltages (Figure (d)). At low voltages, longer spinning distance brought about thinner fiber formation, whereas at high voltages, the effect of spinning distance was totally reversed and fibers with thicker diameters were obtained at longer distances. It is supposed that at low voltages, the stretching time becomes the dominant factor. Hence, longer spinning distance, which gives more time for jet stretching and thinning and solvent evaporation, will result in fibers with smaller diameters. At high voltages, however, the electric field strength is high and dominant. Therefore, increasing the distance, which reduces the electric field, causes an increase in fiber diameter. The function of spinning distance was observed to be independent from volume flow rate for MFD (Figure (e)). The interaction of spinning distance with solution concentration and applied voltage demonstrated in Figure (a) and (d), proved the existence of terms Cd and dV in the model of MFD.

Applied Voltage

Figure 3 (b), (d), and (f) show the effect of applied voltage on MFD. Increasing the voltage resulted in an increase followed by a decrease in MFD. Applied voltage has two major different effects on fiber diameter. Firstly, increasing the applied voltage will increase the electric field strength and larger electrostatic stretching force causes the jet to accelerate more in the electric field, thereby favoring thinner fiber formation. Secondly, since charge transport is only carried out by the flow of polymer in the electrospinning process [72], increasing the voltage would induce more surface charges on the jet. Subsequently, the mass flow rate from the needle tip to the collector will increase, say the solution will be drawn more quickly from the tip of the needle causing fiber diameter to increase. Combination of these two effects will determine the final fiber diameter. Hence, increasing applied voltage may decrease [73]-[75], increase [63], [64], [68] or may not change [25], [31], [69], [76] the fiber diameter. According to the given explanation, at low voltages, where the electric field strength is low, the effect of mass of solution could be dominant. Therefore, fiber diameter increases when the applied voltage rises. However, as the voltage exceeds a limit, the electric field will be high enough to be a determining factor. Hence, fiber diameter decreases as the voltage increases. The effect of voltage on MFD was influenced by solution concentration to some extent (Figure 3 (b)). At high concentrations, the increase in fiber diameter with voltage was more pronounced. This could be attributed to the fact that the effect of mass of solution will be more important for the solutions of higher concentrations. The change in fiber diameter as a function of voltage is dramatically influenced by spinning distance (Figure 3 (d)). At a short distance, the electric field is high and dominant factor. Therefore, increasing applied voltage, which strengthens the electric field, results in a decrease in fiber diameter. Whereas, at long distances where the electric field is low, the effect of mass of solution would be determining factor according to which fiber diameter increased with applied voltage. The effect of applied voltage on MFD is found to be independent from volume flow rate (Figure 3 (f)). It is quite apparent that there is a huge interaction between applied voltage and spinning distance, a slight interaction between applied voltage and solution concentration and no interaction between applied voltage and volume flow rate which is in agreement with the presence of CV and dV and absence of VQ in the model of MFD.

Volume Flow Rate

It was suggested that a minimum value for solution flow rate is required to form the drop of polymer at the tip of the needle for the sake of maintaining a stable Taylor cone [77]. Hence, flow rate could affect the morphology of electrospun nanofibers such as fiber diameter. Increasing the flow rate, more amount of solution is delivered to the tip of the needle enabling the jet to carry the solution away faster. This could bring about an increase in the jet diameter favoring thicker fiber formation. In this study, the MFD slightly increased with volume flow rate (Figure 3 (c), (e), and (f)) which agrees with the previous researches [31], [77]-[79]. Flow rate was also found to influence MFD independent from solution concentration, applied voltage, and spinning distance as suggested earlier by the absence of CQ, dQ, and VQ in the model of MFD.

Response Surfaces for Standard Deviation of Fiber Diameter

Solution Concentration

As depicted in Figure 4 (a), (b), and (c), StdFD increased with concentration which is in agreement with the previous observations [25], [34], [63], [66], [31], [68], [80], [81]. Increasing the polymer concentration, the macromolecular chain entanglements increase, prompting a greater difficulty for the jet to stretch and split. This could result in less uniform fibers (higher StdFD). Concentration affected StdFD regardless of spinning distance (Figure 4 (a)), suggesting that there was no interaction between these two parameters (absence of *Cd* in the model of StdFd). At low applied voltages, the formation of more uniform fibers with decreasing the concentration was facilitated. In agreement with existence of the term *CV* in the model of StdFd, solution concentration was found to have a slight interaction with applied voltage (Figure 4 (b)). The curvature of the surface in Figure 4 (c) suggested that there was a noticeable interaction between concentration and flow rate and this agrees with the presence of the term *CQ* in the model of StdFD.

Spinning Distance

More uniform fibers (lower StdFD) were obtained with increasing the spinning distance as shown in Figure 4 (a), (d), and (e). At longer spinning distance, more time is provided for jet flying from the tip of the needle to the collector and solvent evaporation. Therefore, jet stretching and solvent evaporation is carried out more gently resulting in more uniform fibers. Our finding is consistent with the trend observed by Zhao et al. [81]. Spinning distance influenced StdFD regardless of solution concentration and applied voltage (Figure 4 (a) and (d)) indicating that no interaction exists between these variables as could be inferred from the model of StdFD. However, the interaction of spinning distance with volume flow rate is obvious (Figure 4 (e)). The presence of *dQ* in the model of StdFD proves this observation. The effect of spinning distance is more highlighted at higher flow rates. This could be attributed to the fact that more amount of solution is delivered to the tip of the needle at higher flow rates; therefore the threads will require more time to dry. If the distance is high enough to provide the sufficient time, uniform fibers will be formed. Decreasing the distance, there will be less time for solvent to evaporate favoring the production of non-uniform fibers (high StdFD).

Applied Voltage

StdFD was found to increase with applied voltage (Figure 4 (b), (d), and (f)) as observed in other studies [63], [64], [68], [81]. Increasing the applied voltage causes the effect of the electric field on the charged jet to increase. Hence, the flight speed of the jet increases, shortening the time that the jet travels towards the collector. As a result, less time is provided for jet stretching and thinning and also solvent evaporation. This may result in formation of less uniform fibers (higher StdFD). The effect of applied voltage on StdFD is influenced by solution concentration as shown in Figure 4 (b), implying the interaction of voltage with concentration which was earlier addressed in the paper by the presence of the corresponding term in the model of StdFD. At low concentrations, the formation of uniform fibers with decreasing the applied voltage was facilitated. No interaction was observed between applied voltage and spinning distance (Figure 4 (d)) as suggested by the absence of the term *dV* in the model of StdFD. Figure 4 (f) shows a slight interaction of voltage with flow rate, which concurs with the existence of *VQ* in the model of StdFD.

Volume Flow Rate

As demonstrated in Figure 4 (c), (e), and (f), the uniformity of fibers increased (StdFD decreased), reached to an optimum value and then decreased (StdFD increased) by increasing of the flow rate. When the flow rate is low, the amount of solution fed to the tip of the needle is not sufficient, whereas an excess amount of solution is delivered to the tip of the needle at high flow rates. Therefore, unstable jets are formed in the two extremes resulting in the production of non-uniform fibers. The impact of flow rate on StdFD is influenced by solution concentration, applied voltage, and spinning distance. This observation indicates the interaction between flow rate and other variables as demonstrated by the terms *CQ*, *dQ*, and *VQ* in the model of StdFD. Increasing the solution concentration favored the formation of non-uniform fibers at high flow rates (Figure 4 (c)) which is probably the outcome of greater difficulty of solution removal. The effect of flow rate on StdFD was more pronounced as the spinning distance decreased (Figure 4 (e)). The shorter the distance, the fewer the time provided to the jet to thin and dry. Therefore, at high flow rates, more amount of solution is delivered with insufficient flying time, results in formation of less uniform fibers. High applied voltage encouraged the increase in StdFD at fast flow rates as depicted in Figure 4 (f).

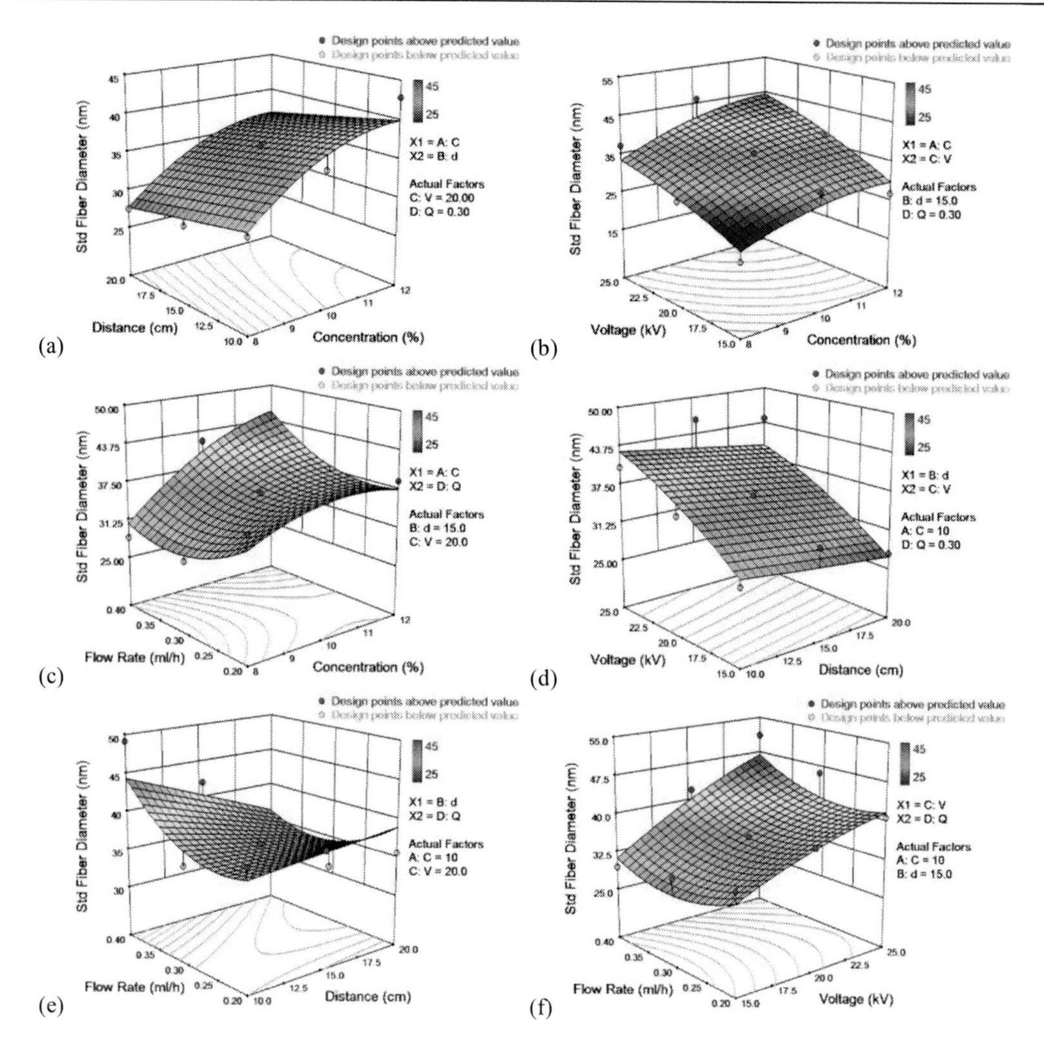

Figure 45. Response surfaces for standard deviation of fiber diameter in terms of: (a) solution concentration and spinning distance, (b) solution concentration and applied voltage, (c) solution concentration and flow rate, (d) spinning distance and applied voltage, (e) spinning distance and flow rate, (f) applied voltage and flow rate.

CONCLUSION

The simultaneous effects of four processing variables including solution concentration, applied voltage, spinning distance, and volume flow rate on MFD and StdFD were investigated quantitatively and qualitatively. The appropriate range of parameters where dry, bead-free, and continuous fibers without breaking up to droplets are formed, were selected by referring to the literature along with conducting a series of preliminary experiments. A full factorial experimental design at three levels of each factor (3^4 design) was carried out. Moreover, 15 treatments inside the design space were selected as test set for evaluating the prediction ability of the models. PVA Nanofibers were then prepared for experimental and test sets through the electrospinning method. After that, MFD and StdFD were determined from SEM micrograph of each sample. RSM was used to establish quadratic models for MFD and StdFD. The test for significance of the coefficients demonstrated that the terms Q^2, CQ, dQ, and VQ in the model of MFD and d^2, Cd, and dV in the model of StdFD were not of important value in representing the responses. Eliminating these terms, simpler yet more efficient models were obtained which not only explained the experimental data in a better manner, but also had more prediction ability. Afterwards, in order to show the generalization ability of the models for predicting new conditions, the test set was used. Low RMSE of test set for MFD and StdFD were obtained indicating the good prediction ability of the models. Finally, in order to qualitatively study the effects of variables on MFD and StdFD, response surface plots were generated using the obtained relationships. For MFD:

1. Increasing solution concentration, MFD increased rigorously. The effect of concentration was more pronounced at longer spinning distance and also at higher applied voltage.
2. The effect of spinning distance on MFD changed depending on solution concentration and applied voltage. At low applied voltages, MFD decreased as the spinning distance became longer, whereas higher MFD resulted with lengthening the spinning distance when the applied voltage was high. Increasing the solution concentration tended to assist the formation of thicker fibers at longer spinning distance.
3. Rising the applied voltage, MFD was observed to first increase and then decrease. High solution concentrations partly and long spinning distances largely favored the increase of MFD with applied voltage.
4. MFD slightly increased with flow rate. The impact of flow rate on MFD was unrelated to the other variables.

For StdFD:

1. The higher the solution concentration, the less uniform fibers (higher StdFD) was formed. Low applied voltages facilitated the formation of more uniform fibers (lower StdFD) with decreasing the concentration. The increase of StdFD with concentration gained momentum at high flow rates.
2. Longer spinning distance resulted in more uniform fibers (lower StdFD). The effect of spinning distance was more pronounced at higher flow rates.
3. Rising the applied voltage increased StdFD. Low concentrations facilitated the formation of uniform fibers (high StdFD) with decreasing the applied voltage.
4. Flow rate was found to have a significant impact on uniformity of fibers (StdFD). As flow rate increased, StdFD decreased and then increased. Higher solution concentration, higher applied voltage, and shorter spinning distance encouraged the formation of non-uniform fibers (high StdFD) at fast flow rates.

REFERENCES

[1] G.I. Taylor, *Proc. Roy. Soc. London.*, 313, 453 (1969).

[2] J. Doshi and D.H. Reneker, *J. Electrostatics*, 35, 151 (1995).

[3] H. Fong and D.H. Reneker, *Electrospinning and the formation of nanofibers*, in: D.R. Salem (Ed.), *Structure formation in polymeric fibers*, Hanser, Cincinnati (2001).

[4] D. Li and Y. Xia, *Adv. Mater.*, 16, 1151 (2004).

[5] R. Derch, A. Greiner and J.H. Wendorff, *Polymer nanofibers prepared by electrospinning*, in: J.A. Schwarz, C.I. Contescu and K. Putyera (Eds.), *Dekker encyclopedia of nanoscience and nanotechnology*, CRC, New York (2004).

[6] A.K. Haghi and M. Akbari, *Phys. Stat. Sol. A*, 204, 1830 (2007).

[7] P.W. Gibson, H.L. Schreuder-Gibson and D. Rivin, *AIChE J.*, 45, 190 (1999).

[8] M. Ziabari, V. Mottaghitalab and A.K. Haghi, *Korean J. Chem. Eng.*, 25, 923 (2008).

[9] Z.M. Huang, Y.Z. Zhang, M. Kotaki and S. Ramakrishna, *Compos. Sci. Technol.*, 63, 2223 (2003).

[10] M. Li, M.J. Mondrinos, M.R. Gandhi, F.K. Ko, A.S. Weiss and P.I. Lelkes, *Biomaterials*, 26, 5999 (2005).

[11] E.D. Boland, B.D. Coleman, C.P. Barnes, D.G. Simpson, G.E. Wnek and G.L. Bowlin, *Acta. Biomater.*, 1, 115 (2005).

[12] J. Lannutti, D. Reneker, T. Ma, D. Tomasko and D. Farson, *Mater. Sci. Eng. C*, 27, 504 (2007).

[13] J. Zeng, L. Yang, Q. Liang, X. Zhang, H. Guan, C. Xu, X. Chen and X. Jing, *J. Control. Release*, 105, 43 (2005).

[14] E.R. Kenawy, G.L. Bowlin, K. Mansfield, J. Layman, D.G. Simpson, E.H. Sanders and G.E. Wnek, *J. Control. Release*, 81, 57 (2002).

[15] M.S. Khil, D.I. Cha, H.-Y. Kim, I.-S. Kim and N. Bhattarai, *J. Biomed. Mater. Res. Part B: Appl. Biomater.*, 67, 675 (2003).

[16] B.M. Min, G. Lee, S.H. Kim, Y.S. Nam, T.S. Lee and W.H. Park, *Biomaterials*, 25, 1289 (2004).

[17] X.H. Qin and S.Y. Wang, *J. Appl. Polym. Sci.*, 102, 1285 (2006).

[18] H.S. Park and Y.O. Park, *Korean J. Chem. Eng.*, 22, 165 (2005).

[19] J.S. Kim and D.H. Reneker, *Poly. Eng. Sci.*, 39, 849 (1999).

[20] S.W. Lee, S.W. Choi, S.M. Jo, B.D. Chin, D.Y. Kim and K.Y. Lee, *J. Power Sources*, 163, 41 (2006).

[21] C. Kim, *J. Power Sources*, 142, 382 (2005).

[22] N.J. Pinto, A.T. Johnson, A.G. MacDiarmid, C.H. Mueller, N. Theofylaktos, D.C. Robinson and F.A. Miranda, *Appl. Phys. Lett.*, 83, 4244 (2003).

[23] D. Aussawasathien, J.-H. Dong and L. Dai, *Synthetic Met.*, 54, 37 (2005).

[24] S.-Y. Jang, V. Seshadri, M.-S. Khil, A. Kumar, M. Marquez, P.T. Mather and G.A. Sotzing, *Adv. Mater.*, 17, 2177 (2005).

[25] S.-H. Tan, R. Inai, M. Kotaki and R. Ramakrishna, *Polymer*, 46, 6128 (2005).

[26] A, Ziabicki, *Fundamentals of fiber formation: The science of fiber spinning and drawing*, Wiley, New York (1976).

[27] A. Podgóski, A. Bałazy and L. Gradoń, *Chem. Eng. Sci.*, 61, 6804 (2006).

[28] B. Ding, M. Yamazaki and S. Shiratori, *Sens. Actuators B*, 106, 477 (2005).

[29] J.R. Kim, S.W. Choi, S.M. Jo, W.S. Lee and B.C. Kim, *Electrochim. Acta*, 50, 69 (2004).
[30] L. Moroni, R. Licht, J. de Boer, J.R. de Wijn and C.A. van Blitterswijk, *Biomaterials*, 27, 4911 (2006).
[31] T. Wang and S. Kumar, *J. Appl. Polym. Sci.*, 102, 1023 (2006).
[32] W. Cui, X. Li, S. Zhou and J. Weng, *J. Appl. Polym. Sci.*, 103, 3105 (2007).
[33] S. Sukigara, M. Gandhi, J. Ayutsede, M. Micklus and F. Ko, *Polymer*, 45, 3701 (2004).
[34] S.Y. Gu, J. Ren and G.J. Vancso, *Eur. Polym. J.*, 41, 2559 (2005).
[35] S.Y. Gu and J. Ren, *Macromol. Mater. Eng.*, 290, 1097 (2005).
[36] O.S. Yördem, M. Papila and Y.Z. Menceloğlu, *Mater. Design*, 29, 34 (2008).
[37] I. Sakurada, *Polyvinyl Alcohol Fibers*, CRC, New York (1985).
[38] F.L. Marten, *Vinyl alcohol polymers*, in: H.F. Mark (Ed.), *Encyclopedia of polymer science and technology*, 3rd ed., vol. 8, Wiley (2004).
[39] Y.D. Kwon, S. Kavesh and D.C. Prevorsek, US Patent, 4,440,711 (1984).
[40] [S. Kavesh and D.C. Prevorsek, US Patent, 4,551,296 (1985).
[41] H. Tanaka, M. Suzuki and F. Uedo, US Patent, 4,603,083 (1986).
[42] G. Paradossi, F. Cavalieri, E. Chiessim, C. Spagnoli and M.K. Cowman, *J. Mater. Sci.: Mater. Med.*, 14, 687 (2003).
[43] G. Zheng-Qiu, X. Jiu-Mei and Z. Xiang-Hong, *Biomed. Mater. Eng.*, 8, 75 (1998).
[44] M. Oka, K. Ushio, P. Kumar, K. Ikeuchi, S.H. Hyon, T. Nakamura and H. Fujita, *P. I. Mech. Eng. H*, 214, 59 (2000).
[45] K. Burczak, E. Gamian and A. Kochman, *Biomaterials*, 17, 2351 (1996).
[46] J.K. Li, N. Wang and X.S. Wu, *J. Control. Release*, 56, 117 (1998).
[47] A.S. Hoffman, *Adv. Drug Delivery Rev.*, 43, 3 (2002).
[48] J. Zeng, A. Aigner, F. Czubayko, T. Kissel, J.H. Wendorff and A. Greiner, *Biomacromolecules*, 6, 1484 (2005).
[49] K.H. Hong, *Polym. Eng. Sci.*, 47, 43 (2007).
[50] M. Ziabari, V. Mottaghitalab and A.K. Haghi, *Korean J. Chem. Eng.*, 25, 919 (2008).
[51] M. Ziabari, V. Mottaghitalab and A.K. Haghi, *Braz. J. Chem. Eng.*, 26, 53 (2009).
[52] M. Ziabari, V. Mottaghitalab, S.T. McGovern and A.K. Haghi, *Nanoscale Res. Lett.*, 2, 597 (2007).
[53] M. Ziabari, V. Mottaghitalab and A.K. Haghi, *Korean J. Chem. Eng.*, 25, 905 (2008).
[54] L.H. Sperling, *Introduction to physical polymer science*, 4th ed., Wiley, New Jersey (2006).
[55] J.C.J.F. Tacx, H.M. Schoffeleers, A.G.M. Brands and L. Teuwen, *Polymer*, 41, 947 (2000).
[56] F.K. Ko, *Nanofiber technology*, in: Y. Gogotsi (Ed.), *Nanomaterials handbook*, CRC, Boca Raton (2006).
[57] A. Koski, K. Yim and S. Shivkumar, *Mater. Lett.*, 58, 493 (2004).
[58] D.C. Montgomery, *Design and analysis of experiments*, 5th ed., Wiley, New York (1997).
[59] A. Dean and D. Voss, *Design and analysis of experiments*, Springer, New York (1999).
[60] G.E.P. Box and N.R. Draper, *Response surfaces, mixtures, and ridge analyses*, Wiley, New Jersey (2007).
[61] K.M. Carley, N.Y. Kamneva and J. Reminga, Response surface methodology, *CASOS Technical Report*, CMU-ISRI-04-136 (2004).

[62] [S. Weisberg, *Applied linear regression*, 3rd ed., Wiley, New Jersey (2005).
[63] C. Zhang, X. Yuan, L. Wu, Y. Han and J. Sheng, *Eur. Polym. J.*, 41, 423 (2005).
[64] Q. Li, Z. Jia, Y. Yang, L. Wang and Z. Guan, Preparation and properties of poly(vinyl alcohol) nanofibers by electrospinning, *Proceedings of IEEE International Conference on Solid Dielectrics*, Winchester, U.K. (2007).
[65] C. Mit-uppatham, M. Nithitanakul and P. Supaphol, *Macromol. Chem. Phys.*, 205, 2327 (2004).
[66] Y.J. Ryu, H.Y. Kim, K.H. Lee, H.C. Park and D.R. Lee, *Eur. Polym. J.*, 39, 1883 (2003).
[67] T. Jarusuwannapoom, W. Hongrojjanawiwat, S. Jitjaicham, L. Wannatong, M. Nithitanakul, C. Pattamaprom, P. Koombhongse, R. Rangkupan and P. Supaphol, *Eur. Polym. J.*, 41, 409 (2005).
[68] S.C. Baker, N. Atkin, P.A. Gunning, N. Granville, K. Wilson, D. Wilson and J. Southgate, *Biomaterials*, 27, 3136 (2006).
[69] S. Sukigara, M. Gandhi, J. Ayutsede, M. Micklus and F. Ko, *Polymer*, 44, 5721 (2003).
[70] X. Yuan, Y. Zhang, C. Dong and J. Sheng, *Polym. Int.*, 53, 1704 (2004).
[71] C.S. Ki, D.H. Baek, K.D. Gang, K.H. Lee, I.C. Um and Y.H. Park, *Polymer*, 46, 5094 (2005).
[72] J.M. Deitzel, J. Kleinmeyer, D. Harris and N.C. Beck Tan, *Polymer*, 42, 261 (2001).
[73] C.J. Buchko, L.C. Chen, Y. Shen and D.C. Martin, *Polymer*, 40, 7397 (1999).
[74] J.S. Lee, K.H. Choi, H.D. Ghim, S.S. Kim, D.H. Chun, H.Y. Kim and W.S. Lyoo, *J. Appl. Polym. Sci.*, 93, 1638 (2004).
[75] S.F. Fennessey and R.J. Farris, *Polymer*, 45, 4217 (2004).
[76] S. Kidoaki, I. K. Kwon and T. Matsuda, *Biomaterials*, 26, 37 (2005).
[77] X. Zong, K. Kim, D. Fang, S. Ran, B.S. Hsiao and B. Chu, *Polymer*, 43, 4403 (2002).
[78] D. Li and Y. Xia, *Nano. Lett.*, 3, 555 (2003).
[79] W.-Z. Jin, H.-W. Duan, Y.-J. Zhang and F.-F. Li, Nonafiber membrane of EVOH-based ionomer by electrospinning, *Proceedings of the 1st IEEE International Conference on Nano/Micro Engineered and Molecular Systems*, Zhuhai, China (2006).
[80] X.M. Mo, C.Y. Xu, M. Kotaki and S. Ramakrishna, *Biomaterials*, 25, 1883 (2004).
[81] S. Zhao, X. Wu, L. Wang and Y. Huang, *J. Appl. Polym. Sci.*, 91, 242 (2004).

APPENDIX

Table 8. Natural and coded variables for experimental and test data along with corresponding responses

No.	Natural Variables				Coded Variables				Responses	
	C (%)	*d* (*cm*)	*V* (*kV*)	*Q* (*ml*/*h*)	x_1	x_2	x_3	x_4	MFD (*nm*)	StdFD (*nm*)
1	8	10	15	0.2	-1	-1	-1	-1	232.62	26.60
2	8	10	15	0.3	-1	-1	-1	0	235.50	24.52
3	8	10	15	0.4	-1	-1	-1	1	252.02	25.89
4	8	10	20	0.2	-1	-1	0	-1	236.84	37.30
5	8	10	20	0.3	-1	-1	0	0	232.08	30.22
6	8	10	20	0.4	-1	-1	0	1	249.21	34.49

Table 8. (continued)

No.	Natural Variables				Coded Variables				Responses	
	C (%)	d (cm)	V (kV)	Q (ml/h)	x_1	x_2	x_3	x_4	MFD (nm)	StdFD (nm)
7	8	10	25	0.2	-1	-1	1	-1	196.05	34.76
8	8	10	25	0.3	-1	-1	1	0	201.38	35.15
9	8	10	25	0.4	-1	-1	1	1	215.00	39.00
10	8	15	15	0.2	-1	0	-1	-1	221.10	28.88
11	8	15	15	0.3	-1	0	-1	0	238.63	20.17
12	8	15	15	0.4	-1	0	-1	1	242.32	21.99
13	8	15	20	0.2	-1	0	0	-1	219.76	36.19
14	8	15	20	0.3	-1	0	0	0	228.56	28.29
15	8	15	20	0.4	-1	0	0	1	242.01	28.30
16	8	15	25	0.2	-1	0	1	-1	202.62	33.22
17	8	15	25	0.3	-1	0	1	0	208.21	37.14
18	8	15	25	0.4	-1	0	1	1	213.66	34.84
19	8	20	15	0.2	-1	1	-1	-1	196.63	30.69
20	8	20	15	0.3	-1	1	-1	0	197.73	24.55
21	8	20	15	0.4	-1	1	-1	1	206.28	22.11
22	8	20	20	0.2	-1	1	0	-1	206.69	31.56
23	8	20	20	0.3	-1	1	0	0	224.38	27.41
24	8	20	20	0.4	-1	1	0	1	242.06	26.51
25	8	20	25	0.2	-1	1	1	-1	205.25	40.32
26	8	20	25	0.3	-1	1	1	0	215.70	30.54
27	8	20	25	0.4	-1	1	1	1	231.34	32.40
28	10	10	15	0.2	0	-1	-1	-1	269.91	30.35
29	10	10	15	0.3	0	-1	-1	0	270.05	28.88
30	10	10	15	0.4	0	-1	-1	1	291.99	33.98
31	10	10	20	0.2	0	-1	0	-1	256.11	38.54
32	10	10	20	0.3	0	-1	0	0	264.86	35.70
33	10	10	20	0.4	0	-1	0	1	278.34	49.13
34	10	10	25	0.2	0	-1	1	-1	228.21	42.33
35	10	10	25	0.3	0	-1	1	0	239.28	40.30
36	10	10	25	0.4	0	-1	1	1	238.74	46.57
37	10	15	15	0.2	0	0	-1	-1	263.67	34.16
38	10	15	15	0.3	0	0	-1	0	269.29	31.54
39	10	15	15	0.4	0	0	-1	1	277.71	29.40
40	10	15	20	0.2	0	0	0	-1	284.20	38.18
41	10	15	20	0.3	0	0	0	0	281.82	36.27
42	10	15	20	0.4	0	0	0	1	282.39	42.07
43	10	15	25	0.2	0	0	1	-1	249.42	40.79
44	10	15	25	0.3	0	0	1	0	278.22	46.15
45	10	15	25	0.4	0	0	1	1	286.96	51.16
46	10	20	15	0.2	0	1	-1	-1	239.45	27.98
47	10	20	15	0.3	0	1	-1	0	244.04	27.43
48	10	20	15	0.4	0	1	-1	1	251.58	27.26
49	10	20	20	0.2	0	1	0	-1	285.67	35.62
50	10	20	20	0.3	0	1	0	0	273.05	30.74
51	10	20	20	0.4	0	1	0	1	280.62	34.66
52	10	20	25	0.2	0	1	1	-1	278.10	40.79
53	10	20	25	0.3	0	1	1	0	280.95	44.58
54	10	20	25	0.4	0	1	1	1	306.28	44.04

Table 8. (continued)

No.	Natural Variables				Coded Variables				Responses	
	C (%)	d (cm)	V (kV)	Q (ml/h)	x_1	x_2	x_3	x_4	MFD (nm)	StdFD (nm)
55	12	10	15	0.2	1	-1	-1	-1	286.23	27.12
56	12	10	15	0.3	1	-1	-1	0	295.60	32.91
57	12	10	15	0.4	1	-1	-1	1	293.41	40.48
58	12	10	20	0.2	1	-1	0	-1	271.20	34.86
59	12	10	20	0.3	1	-1	0	0	291.89	42.78
60	12	10	20	0.4	1	-1	0	1	295.93	49.43
61	12	10	25	0.2	1	-1	1	-1	234.13	39.31
62	12	10	25	0.3	1	-1	1	0	247.65	48.60
63	12	10	25	0.4	1	-1	1	1	247.13	59.02
64	12	15	15	0.2	1	0	-1	-1	271.93	33.05
65	12	15	15	0.3	1	0	-1	0	297.65	26.75
66	12	15	15	0.4	1	0	-1	1	296.79	39.84
67	12	15	20	0.2	1	0	0	-1	297.94	38.82
68	12	15	20	0.3	1	0	0	0	310.06	36.84
69	12	15	20	0.4	1	0	0	1	312.15	41.69
70	12	15	25	0.2	1	0	1	-1	272.24	39.55
71	12	15	25	0.3	1	0	1	0	282.04	42.35
72	12	15	25	0.4	1	0	1	1	288.00	51.72
73	12	20	15	0.2	1	1	-1	-1	259.63	34.63
74	12	20	15	0.3	1	1	-1	0	278.40	25.35
75	12	20	15	0.4	1	1	-1	1	279.25	27.25
76	12	20	20	0.2	1	1	0	-1	307.42	42.25
77	12	20	20	0.3	1	1	0	0	327.77	35.71
78	12	20	20	0.4	1	1	0	1	337.88	45.16
79	12	20	25	0.2	1	1	1	-1	321.78	46.21
80	12	20	25	0.3	1	1	1	0	334.54	40.68
81	12	20	25	0.4	1	1	1	1	342.45	47.94
82	9	20	15	0.3	-0.5	1	-1	0	216.53	24.25
83	10	12.5	15	0.3	0	-0.5	-1	0	259.61	25.67
84	10	20	22.5	0.3	0	1	0.5	0	300.27	35.71
85	10	20	15	0.25	0	1	-1	-0.5	235.04	29.64
86	9	12.5	15	0.3	-0.5	-0.5	-1	0	247.57	26.65
87	9	20	22.5	0.3	-0.5	1	0.5	0	247.16	31.12
88	9	20	15	0.25	-0.5	1	-1	-0.5	212.82	30.26
89	10	12.5	22.5	0.3	0	-0.5	0.5	0	263.70	45.06
90	10	12.5	15	0.25	0	-0.5	-1	-0.5	258.26	26.16
91	10	20	22.5	0.25	0	1	0.5	-0.5	272.03	36.28
92	9	12.5	22.5	0.3	-0.5	-0.5	0.5	0	235.75	33.16
93	9	12.5	15	0.25	-0.5	-0.5	-1	-0.5	244.43	24.87
94	9	20	22.5	0.25	-0.5	1	0.5	-0.5	252.50	36.01
95	10	12.5	22.5	0.25	0	-0.5	0.5	-0.5	260.71	42.25
96	9	12.5	22.5	0.25	-0.5	-0.5	0.5	-0.5	231.97	32.86

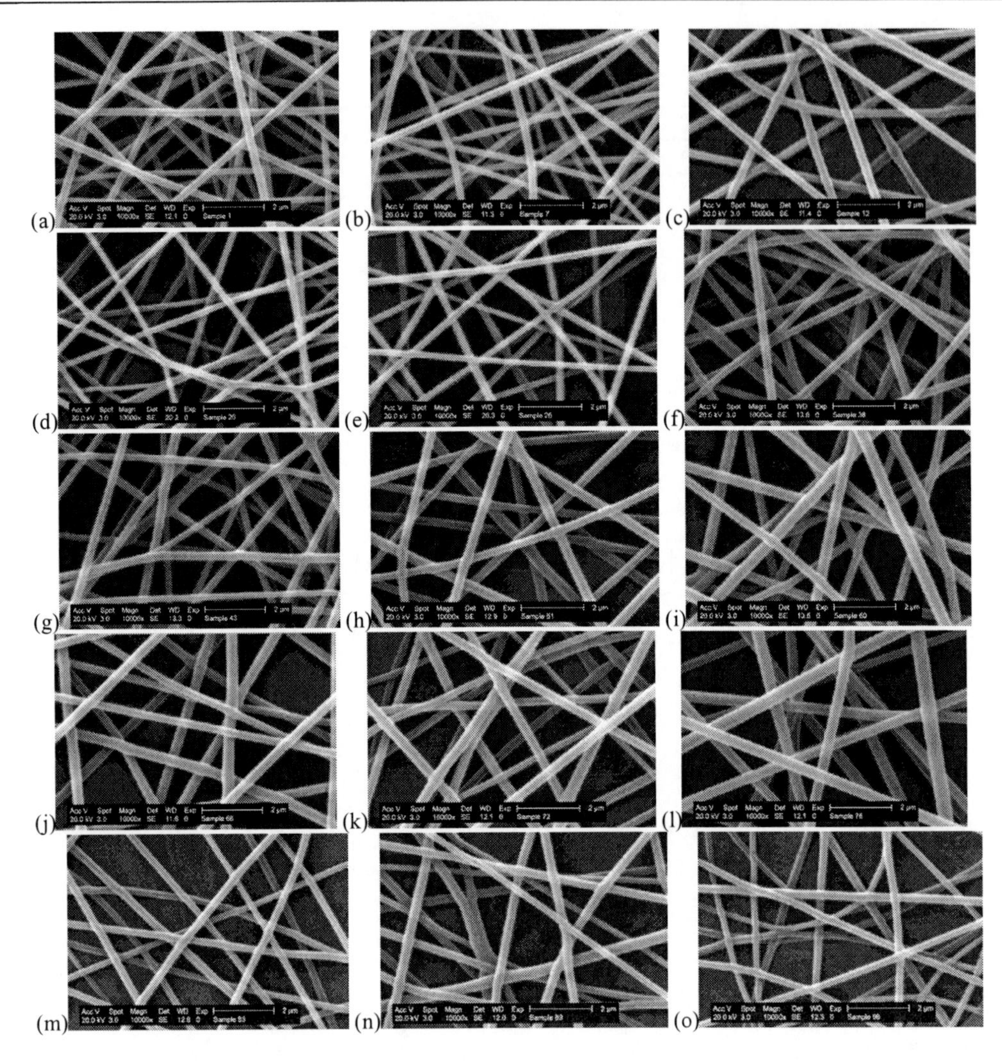

Figure 5. SEM micrographs of typical PVA electrospun nanofiber mats: (a) *C*= 8%, *d*= 10*cm*, *V*= 15*kV*, Q= 0.2*ml*/*h*, (b) *C*= 8%, *d*= 10*cm*, *V*= 25*kV*, Q= 0.2*ml*/*h*, (c) *C*= 8%, *d*= 15*cm*, *V*= 15*kV*, Q= 0.4*ml*/*h*, (d) *C*= 8%, *d*= 20*cm*, *V*= 15*kV*, Q= 0.3*ml*/*h*, (e) *C*= 8%, *d*= 20*cm*, *V*= 25*kV*, Q= 0.3*ml*/*h*, (f) *C*= 10%, *d*= 15*cm*, *V*= 15*kV*, Q= 0.3*ml*/*h*, (g) *C*= 10%, *d*= 15*cm*, *V*= 25*kV*, Q= 0.2*ml*/*h*, (h) *C*= 10%, *d*= 20*cm*, *V*= 20*kV*, Q= 0.4*ml*/*h*, (i) *C*= 12%, *d*= 10*cm*, *V*= 20*kV*, Q= 0.4*ml*/*h*, (j) *C*= 12%, *d*= 15*cm*, *V*= 15*kV*, Q= 0.4*ml*/*h*, (k) *C*= 12%, *d*= 15*cm*, *V*= 25*kV*, Q= 0.4*ml*/*h*, (l) *C*= 12%, *d*= 20*cm*, *V*= 20*kV*, Q= 0.4*ml*/*h*, (m) *C*= 10%, *d*= 12.5*cm*, *V*= 15*kV*, Q= 0.3*ml*/*h*, (n) *C*= 10%, *d*= 12.5*cm*, *V*= 22.5*kV*, Q= 0.3*ml*/*h*, (o) *C*= 9%, *d*= 12.5*cm*, *V*= 22.5*kV*, Q= 0.25*ml*/*h*.

Chapter 4

ELECTROSPUN NANO-FIBER MEASUREMENT IN NONWOVENS

1. INTRODUCTION

In Electrospinning process, a high electric field is generated between a polymer solution held by its surface tension at the end of a syringe (or a capillary tube) and a collection target. Figure 1 illustrates the electrospinning setup.

Accurate and automated measurement of nanofiber diameter is useful and crucial and therefore has been taken into consideration in this contribution. The objective of the current research would then be to develop an image analysis based method to serve as a simple, automated and efficient alternative for electrospun nanofiber diameter measurement.

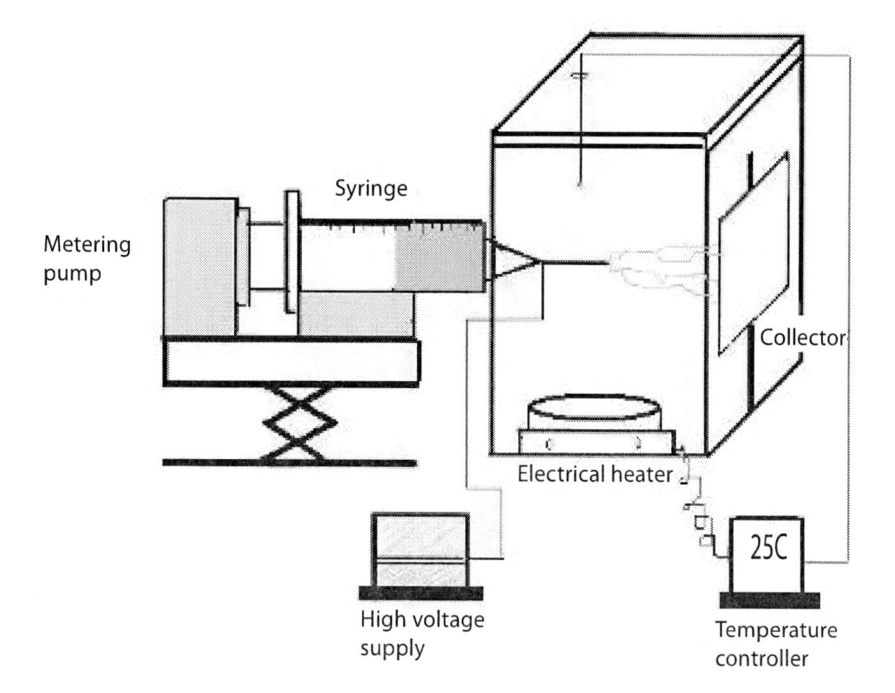

Figure 1. Electrospinning setup.

2. Methodology

The algorithm for determining fiber diameter uses a binary input image and creates its skeleton and distance transformed image (distance map). The skeleton acts as a guide for tracking the distance transformed image and fiber diameters are measured from the intensities of the distance map at all points along the skeleton . Figure 2 shows a simple simulated image, which consists of five fibers with diameters of 10, 13, 16, 19 and 21 pixels, together with its skeleton and distance map including the histogram of fiber diameter obtained by this method.

In this paper, we developed *direct tracking* method for measuring electrospun nanofiber diameter. This method which also uses a binary image as the input, determines fiber diameter based on information acquired from two scans; first a horizontal and then a vertical scan. In the horizontal scan, the algorithm searches for the first white pixel (representative of fibers) adjacent to a black (representative of background). Pixels are then counted until reaching the first black. Afterwards, the second scan is started from the mid point of horizontal scan and pixels are counted until the first vertical black pixel is encountered. Direction will change if the black pixel isn't found (Figure 3). Having the number of horizontal and vertical scans, the number of pixels in perpendicular direction which is the fiber diameter in terms of pixels can be measured through a simple geometrical relationship.

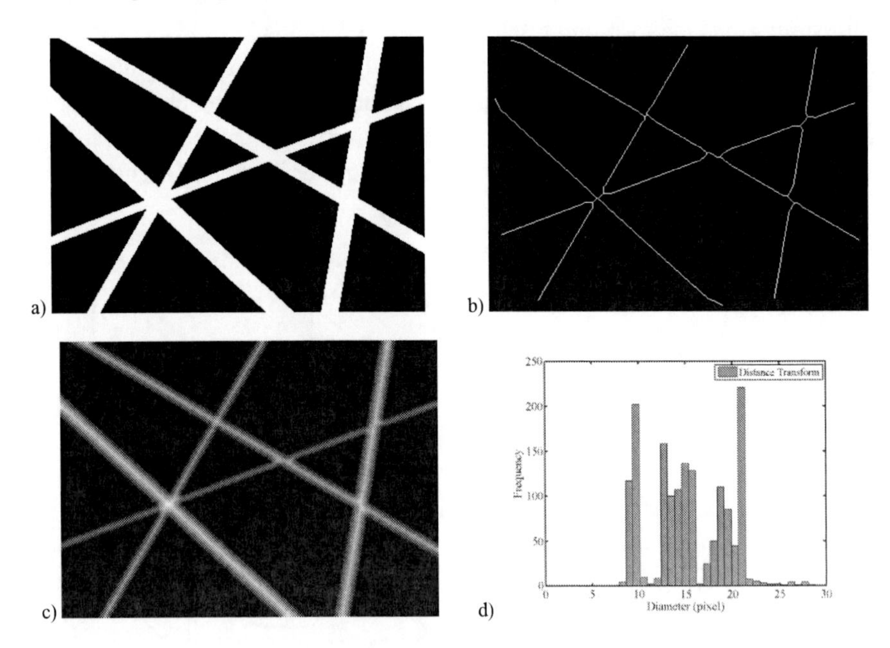

Figure 2. a) A simple simulated image, b) Skeleton of (a), c) Distance map of (a) after pruning, d) Histogram of fiber diameter distribution obtained by distance transform method.

In electrospun webs, nanofibers cross each other at intersection points and this brings about the possibility for some untrue measurements of fiber diameter in these regions. To circumvent this problem, a process called *fiber identification* is employed. First, black regions are labeled and a couple of regions between which a fiber exists, are selected. Figure depicts the labeled simulated image and the histogram of fiber diameter obtained by direct tracking method.

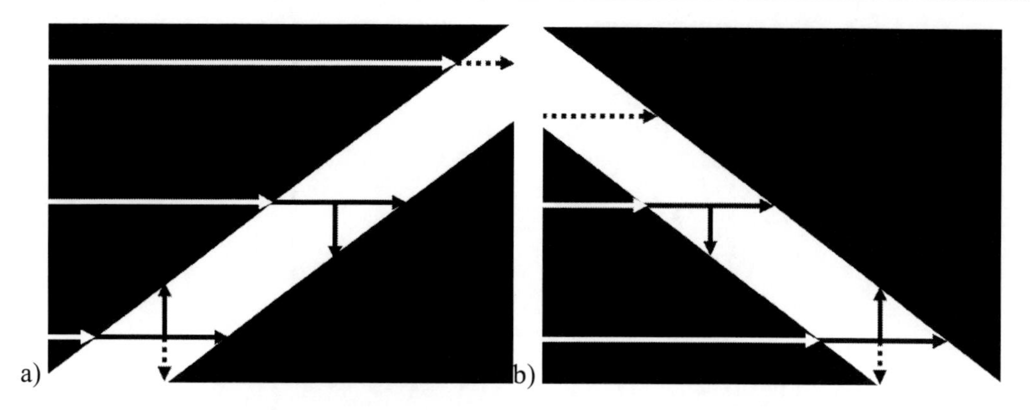

Figure 3. Fiber diameter measurement based on two scans in direct tracking method.

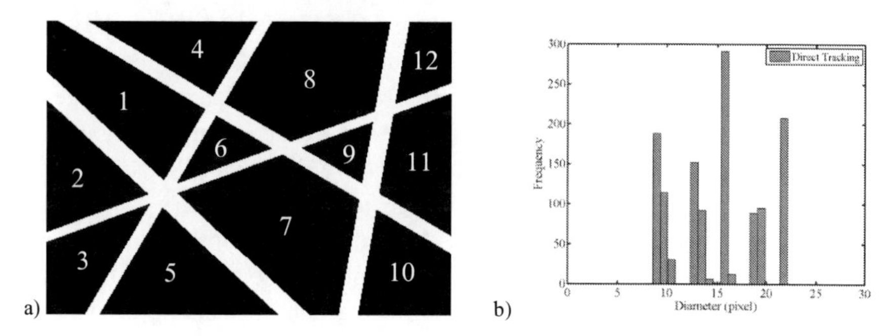

Figure 4. a) The labeled simulated image, b) Histogram of fiber diameter distribution obtained by direct tracking method.

Now, reliable evaluation of the accuracy of the developed methods requires samples with known characteristics. Since it is neither possible to obtain real electrospun webs with specific characteristics through the experiment nor there is a method which measures fiber diameters precisely with which to compare the results, the method will not be well evaluated using just real webs. To that end, a simulation algorithm has been employed for generating samples with known characteristics. In this case, it is assumed that the lines are infinitely long so that in the image plane, they intersect the boundaries. Under this scheme, which is shown in Figure 5, a line with a specified thickness is defined by the perpendicular distance d away from a fixed reference point O located in the center of the image and the angular position of the perpendicular α. Distance d is limited to the diagonal of the image . Several variables are allowed to be controlled during simulation; line thickness, line density, angular density and distance from the reference point. These variables can be sampled from given distributions or held constant.

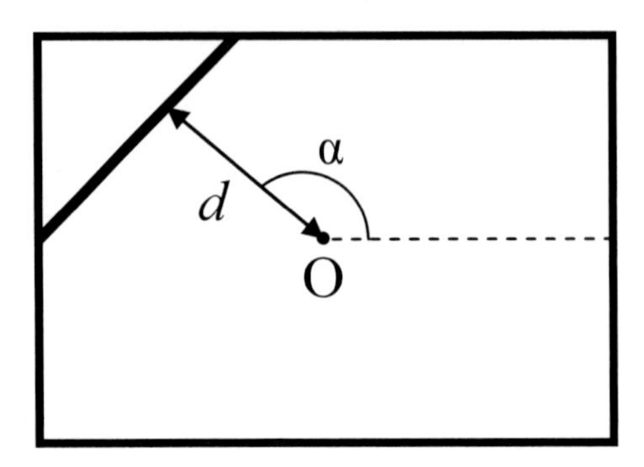

Figure 5. μ-randomness procedure.

Distance transform and direct tracking algorithms for measuring fiber diameter both require binary image as their input. Hence, the micrographs of electrospun webs first have to be converted to black and white. This is carried out by *thresholding* process (known also as *segmentation*) which produces binary image from a grayscale (intensity) image . This is a critical step because the segmentation affects the result significantly. Prior to the segmentation, an *intensity adjustment* operation and a two dimensional *median* filter are often applied in order to enhance the contrast of the image and remove noise.

In the simplest thresholding technique, called *global thresholding*, the image is segmented using a single constant threshold. One simple way to choose a threshold is by trial and error. Each pixel is then labeled as object or background depending on whether its gray level is greater or less than the value of threshold respectively.

The main problem of global thresholding is its possible failure in the presence of non-uniform illumination or local gray level unevenness. An alternative to this problem is to use *local thresholding* instead. In this approach, the original image is divided to subimages and different thresholds are used for segmentation. As it is shown in Figure 6, global thresholding resulted in some broken fiber segments. This problem was solved using local thresholding.

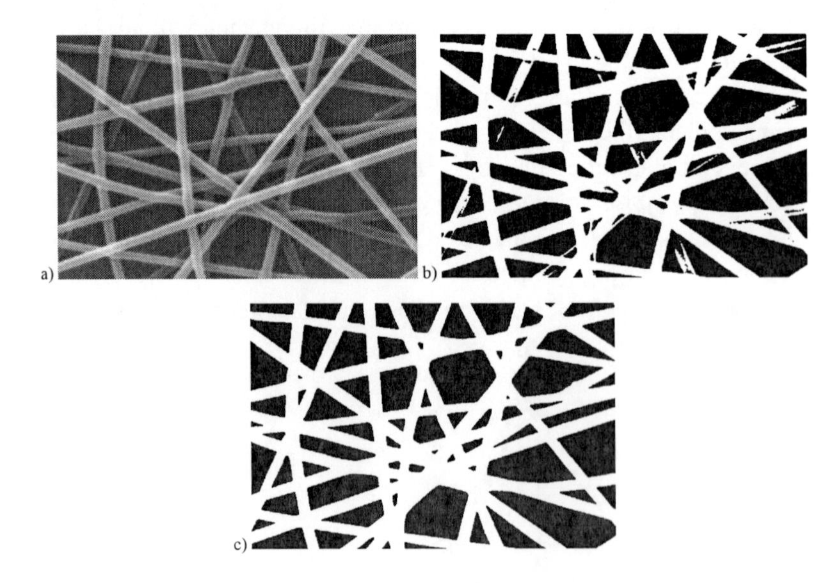

Figure 6. a) A typical electrospun web, b) Global thresholding, c) Local thresholding.

3. EXPERIMENTAL

Electrospun nanofiber webs used as real webs in image analysis were prepared by electrospinning aqueous solutions of PVA with average molecular weight of 72000 *g/mol* (MERCK) at different processing parameters. The micrographs of the webs were obtained using a Philips (XL-30) environmental Scanning Electron Microscope (SEM) under magnification of 10000X after gold sputter coating.

4. RESULTS AND DISCUSSION

Three simulated images generated by μ-randomness procedure were used as samples with known characteristics to demonstrate the validity of the techniques. They were each produced by 30 randomly oriented lines with varied diameters sampled from normal distributions with mean of 15 pixels and standard deviation of 2, 4 and 8 pixels respectively. Table 1 summarizes the structural features of these simulated images which are shown in Figure 7.

Table 1. Structural characteristics of the simulated images generated using μ-randomness procedure

No.	Angular range	Line density	Line thickness	
			M	Std
1	0-360	30	15	2
2	0-360	30	15	4
3	0-360	30	15	8

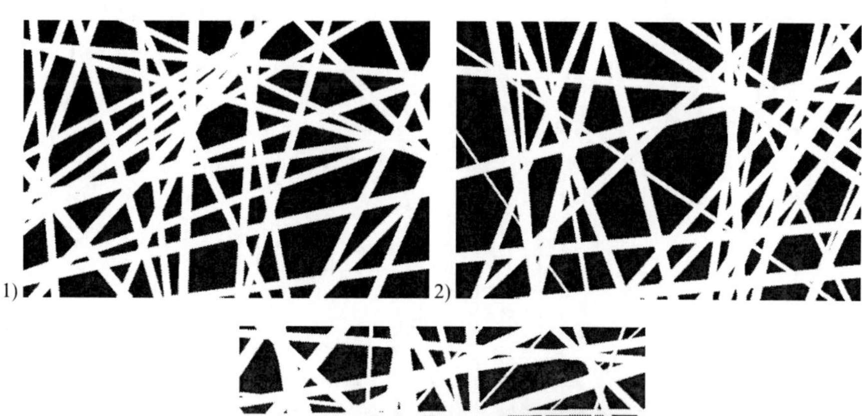

Figure 7. Simulated images generated using μ-randomness procedure.

Mean and standard deviation of fiber diameters for the simulated images obtained by direct tracking as well as distance transform are listed in Table 2. Figure 8 shows histograms of fiber diameter distribution for the simulated images obtained by the two methods. In order to make a true comparison, the original distribution of fiber diameter in each simulated image is also included. The line over each histogram is related to the fitted normal distribution to the corresponding fiber diameters.

Table 2. Mean and standard deviation of fiber diameters for the simulated images

		No. 1	No. 2	No. 3
Simulation	M	15.247	15.350	15.367
	Std	1.998	4.466	8.129
Distance transform	M	16.517	16.593	17.865
	Std	5.350	6.165	9.553
Direct tracking	M	16.075	15.803	16.770
	Std	2.606	5.007	9.319

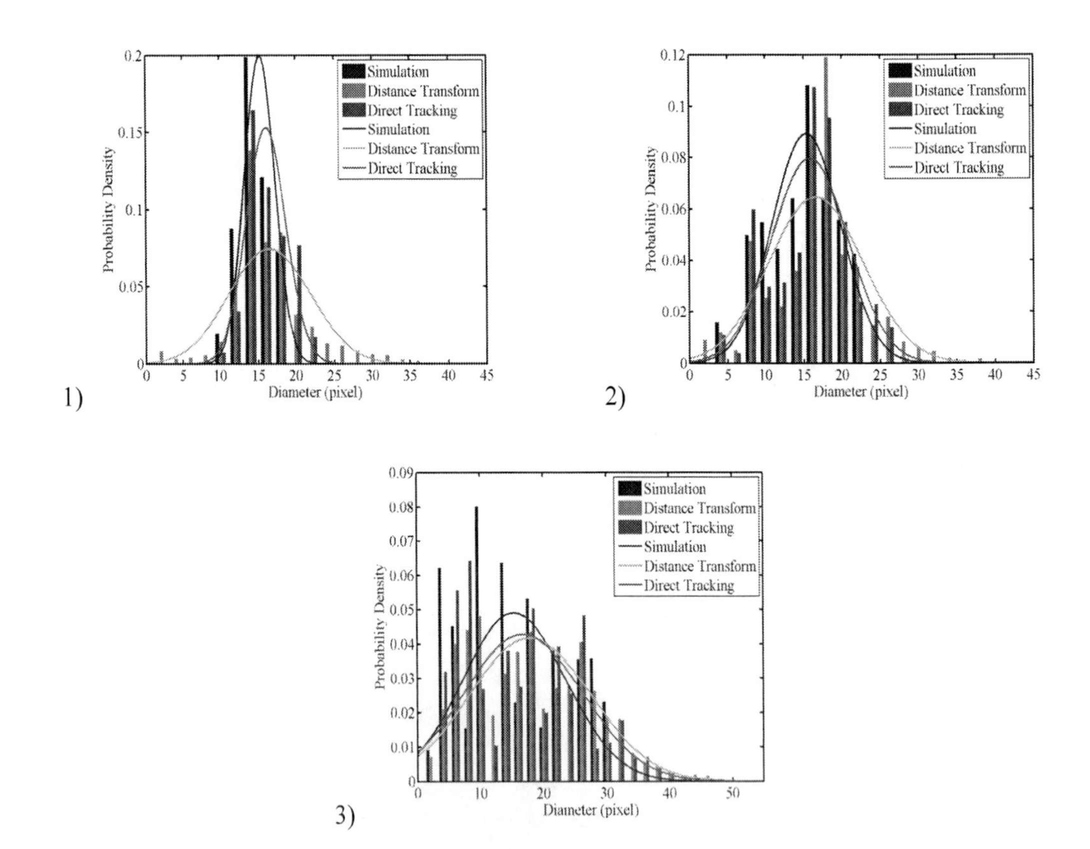

Figure 8. Histograms of fiber diameter distribution for the simulated images.

Table 2 and Figure 8 clearly demonstrate that for all simulated webs, direct tracking method resulted in mean and standard deviation of fiber diameters which are closer to those of the corresponding simulated image (the true ones). Distance transform method is far away from making reliable and accurate measurements. This may be due to remaining some branches in the skeleton even after pruning. The thicker the line, the higher possibility of branching during skeletonization (or thinning). Although these branches are small, their orientation is typically normal to the fiber axis; thus causing widening the distribution obtained by distance transform method.

Conclusion

Fiber diameter is one of the most important structural characteristics in electrospun nanofiber webs. Electrospun nanofiber diameter is often measured by manual method – labor intensive, time consuming, operator-based technique which only utilizes a low number of measurements, thereby inefficient for automated systems e.g. online quality control. In this study, an automated technique called "Direct Tracking" for measuring electrospun nanofiber diameter has been developed that is fast and has the capacity for automation, enabling improved quality control of large scale electrospinning operations.

References

[1] M. Ziabari, V. Mottaghitalab, A.K.Haghi, Application of direct tracking method for measuring electrospun nanofiber diameter. *Braz. J. Chem. Eng.* , v. 26, n. 1, pp. 53-62, 2009

[2] M. Ziabari, V. Mottaghitalab, S. T. McGovern, A. K. Haghi, Measuring Electrospun Nanofibre Diameter: A Novel Approach, *Chin.Phys.Lett.* , Vol. 25, No. 8 , pp. 3071-3074, 2008

[3] M. Ziabari, V. Mottaghitalab, S. T. McGovern, A. K. Haghi, A new image analysis based method for measuring electrospun nanofiber diameter, *Nanoscale Research Letter* , Vol. 2, pp. 297-600, 2007.

[4] M. Ziabari, V. Mottaghitalab, A. K. Haghi, Simulated image of electrospun nonwoven web of PVA and corresponding nanofiber diameter distribution, *Korean Journal of Chemical engng* , Vol.25, No. 4, pp. 919-922 ,2008.

[5] M. Ziabari, V. Mottaghitalab, A. K. Haghi, Evaluation of electrospun nanofiber pore structure parameters, *Korean Journal of Chemical engng* , Vol.25, No. 4, pp. 923-932 ,2008.

[6] M. Ziabari, V. Mottaghitalab, A. K. Haghi, Distance transform algoritm for measuring nanofiber diameter, *Korean Journal of Chemical engng* , Vol25, No. 4, pp. 905-918 ,2008.

[7] A.K.Haghi and M. Akbari, Trends in electrospinning of natural nanofibers, *Physica Status Solidi* , Vol.204, No. 6 pp. 1830–1834 , 2007.

[8] M. Ziabari, V. Mottaghitalab, A. K. Haghi, A new approach for optimization of electrospun nanofiber formation process, *Korean Journal of Chemical engng,* DOI: 10.2478/s11814-009-0309-1

Chapter 5

ELECTROSPUN POLYACRYLONITRILE NANOFIBERS

1. INTRODUCTION

Electrospinning is a novel and efficient method by which fibers with diameters in nanometer scale entitled as nanofibers, can be achieved. In electrospinning process, a strong electric field is applied on a droplet of polymer solution (or melt) held by its surface tension at the tip of a syringe's needle (or a capillary tube). As a result, the pendent drop will become highly electrified and the induced charges are distributed over its surface. Increasing the intensity of electric field, the surface of the liquid drop will be distorted to a conical shape known as the Taylor cone [1-4]. Once the electric field strength exceeds a threshold value, the repulsive electric force dominates the surface tension of the liquid and a stable jet emerges from the cone tip. The charged jet is then accelerated toward the target and rapidly thins and dries as a result of elongation and solvent evaporation. As the jet diameter decreases, the surface charge density increases and the resulting high repulsive forces split the jet to smaller jets. This phenomenon may take place several times leading to many small jets. Ultimately, solidification is carried out and fibers are deposited on the surface of the collector as a randomly oriented nonwoven mat [5-7]. Figure 1 shows a schematic illustration of conventional electrospinning setup and the setup used for this study..

In this study, a simple and non-conventional electrospinning technique was employed for producing of highly oriented Polyacrylonitrile (PAN) monofilament nanofibers. In the present study the electrospinning process was carried out using two needles in opposite positions and a rotating collector perpendicular to needle axis.

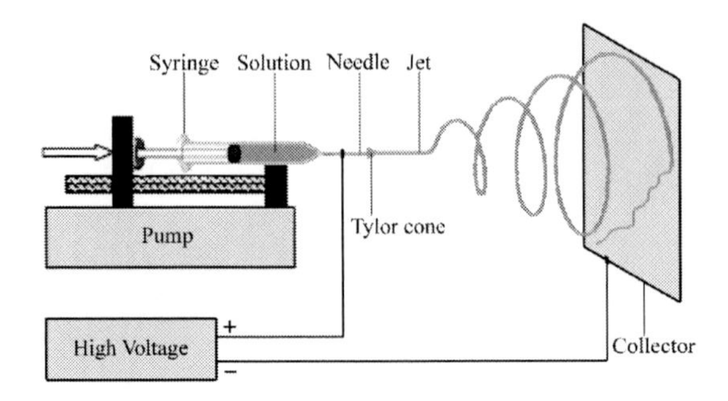

(a) conventional electrospinning technique

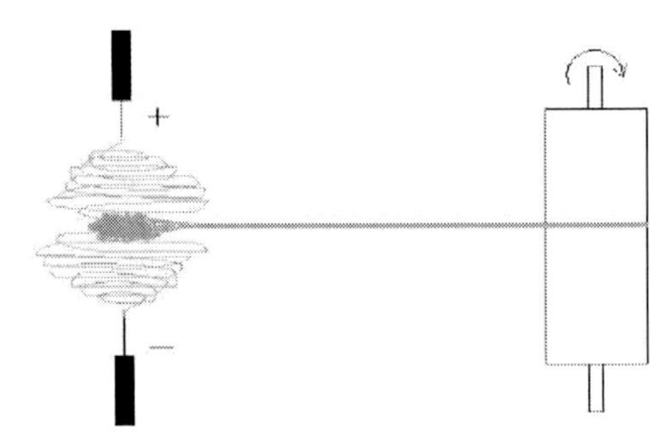

(b) non-conventional electrospinning technique (present study)

Figure 1. Schematic illustration of electrospinning setup: (a) Conventional, (b) Present Study

2. Experimental

The electrospinning apparatus consist of

- a high voltage power supply,
- two syringe pumps,
- two stainless steel needles (0.7 mm OD) and
- a rotating collector with variable surface speed which is controlled by an Inverter. In this setup unlike the conventional technique, two needles were installed in opposite direction and polymer solutions were pumped to needles by two syringe infusion pumps with same feed rate.

3. Results and Discussion

3.1. Productivity

In this study, fibers were electrospun in the aligned form by using a simple and novel method which manufacture well-aligned polymer nanofibers with infinite length and over large collector area [4]. The conditions pertinent to minimum number of rupture for nanofibers prepared at different concentrations are shown in Table 1.

Table 1. The conditions obtained to generate nanofibers with minimum amount of rupture

Solution concentration (Wt%)	Applied voltage (Kv)	Feed rate (mL/h)	Distance between two needles (cm)	Distance between needles and collector (cm)	Total number of rupture (at 15 miniutes)
13	10.5	0.293	13	20	12
14	11	0.293	13	20	6
15	11	0.293	15	20	6

PAN nanofibers prepared at 14 and 15 wt% concentrations have lower rupture which can be due to high chain entanglement in these oncentrations. Fig 3 shows SEM images of PAN nanofibers electrospun in obtained optimum conditions. The average diameter of nanofibers increased by increasing solution concentration (Table 2).

Table 2. Average diameter of electrospun nanofibers at different solution concentrations

Diameter (nm) / Concentration	Average	Coefficient Variation (CV%)	$\overline{x} \pm sd$ (nm)
13 wt%	323.45	9.59	323.45 ± 31.03
14 wt%	394.19	7.32	394.19 ± 28.84
15 wt%	404.67	10.82	404.67 ± 43.81

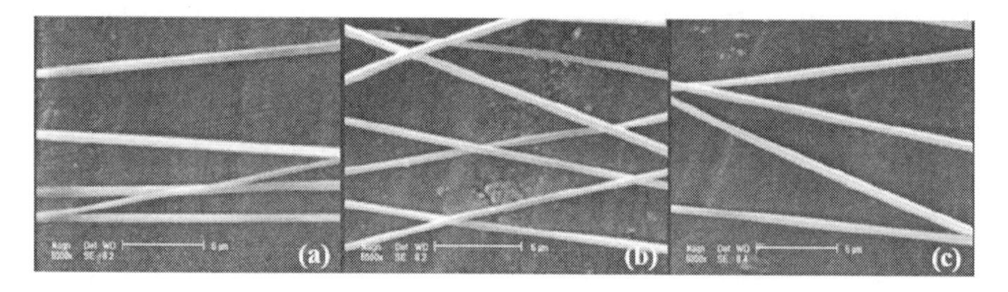

Figure 3. SEM images of PAN nanofibers at concentrations of (a) wt% (b) w% (c) 15 %.

3.2. Alignment

Analysis of fibers alignment was carried out by obtaining angular power spectrum (APS) of nanofibers collected at different take up speeds from 22.5 m/min to 67.7 m/min. The plot of normalized APS (ratio of intensity of the APS to the corresponding mean intensity of the Fourier power spectrum) versus angle was used for calculating degree of alignment (Fig 4). The alignment of the collected fibers is induced by the rotation of the target and improves as the surface velocity of the target is increased (Table 3).

Table 3. The degree of alignment of the collected fibers

Take-up speed (m/min)	22.5	31.6	40.6	49.6	59.5	67.7
Degree of alignment (%)	24.59±3.97	34.4±5.29	32.72±7.65	29.48±5.97	37.53±5.26	29.43±7.04

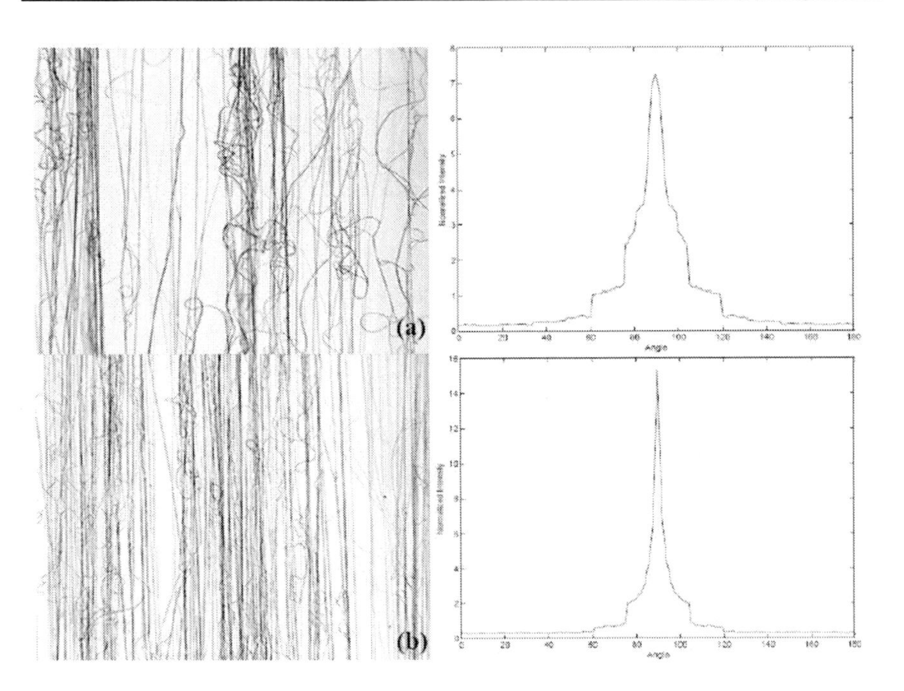

Figure 4. Optical micrograph of electrospun PAN nanofibers with corresponding normalized APS at take-up speeds of (a) 22.5 m/min (b) 59.5 m/min.

3.3. Crystallization Index

The crystallization index (A1730/A2240) was calculated from FTIR spectra of PA
N nanofibers collected at different take-up speeds from 22.5 m/min to 67.7 m/min in optimum conditions (Fig 5).

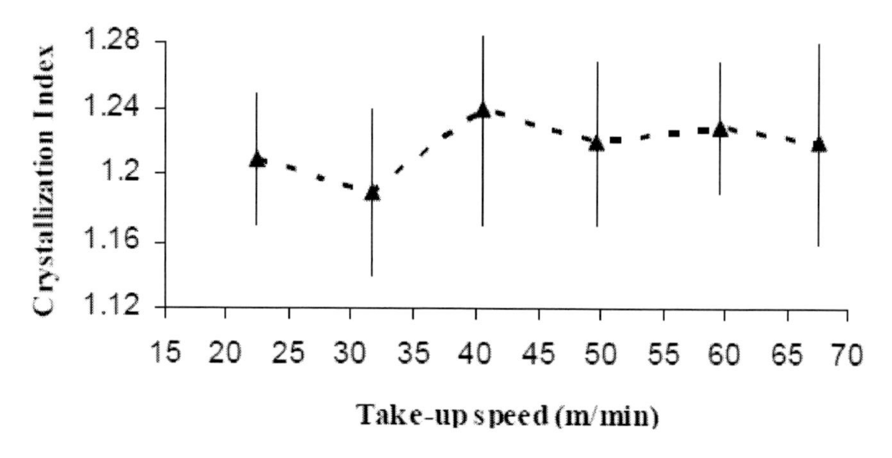

Figure 5. Crystallization Index of PAN nanofibers versus take up speed.

3.4. Molecular Orientation

The molecular orientation of the nanofibers was examined by Raman Spectroscopy. Raman spectra were collected from bundles of fibers electrospun at 11 kV from 14 wt% PAN in DMF solutions collected onto a drum rotating with a surface velocity between 22.5 m/min and 67.7 m/min. The main difference among different molecular structures of PAN fibers usually arise in the region of 500-1500 cm-1 which is called as finger point region [9]. In this region, the peaks over the ranges of 950-1090 cm-1 and 1100-1480 cm-1 are common [9,10] observed at Raman spectra of generated PAN nanofibers (Fig 6).

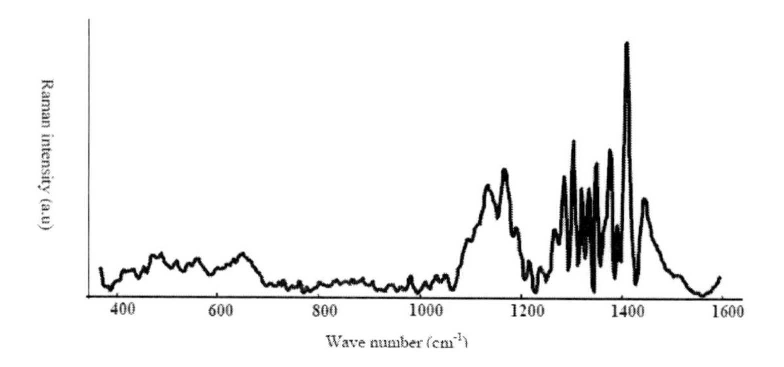

Figure 6. Raman spectra of PAN nanofiber.

Figure 7 shows the Raman spectra of different samples of the nanofibers using VV configuration for different amounts of ψ. Raman spectra were obtained in two directions, parallel (ψ=0°) and vertical direction (ψ=90°) with respect to the polarization plane.

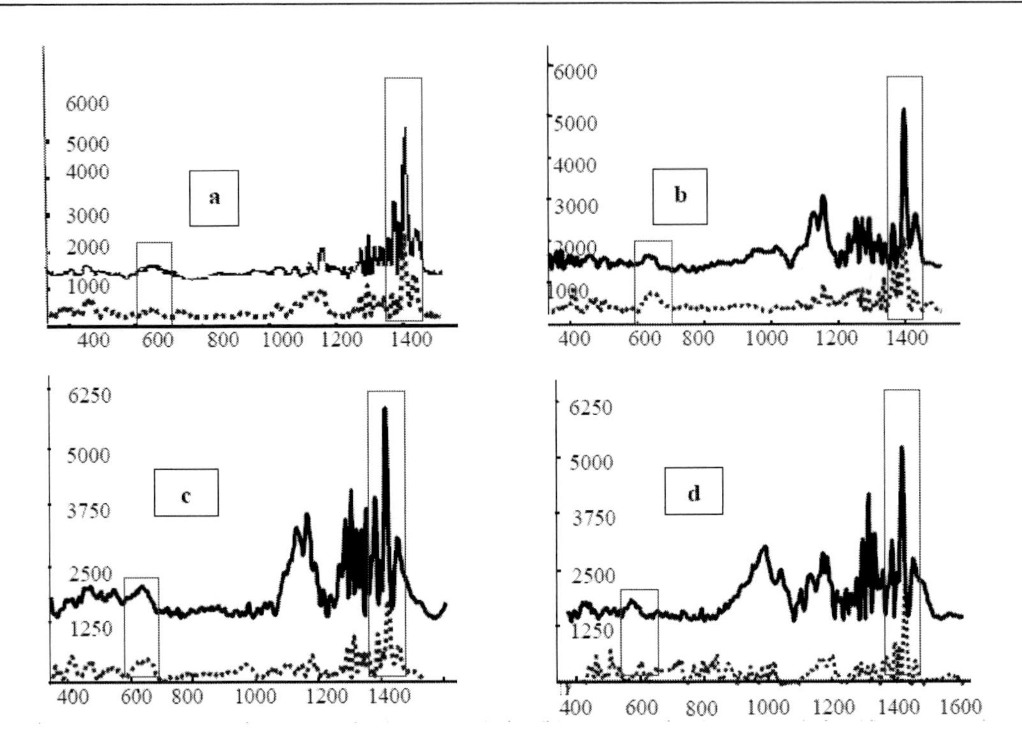

Figure 7. The Raman spectra under VV mode. (a) 22.5 m/min (c) 59.5 m/min (d) 67.7 m/min. From top to bottom the angel between fiber axis and polarization plane if 0° and 90°.

Trend in peak intensity observed for different sample shown in Figure 8.

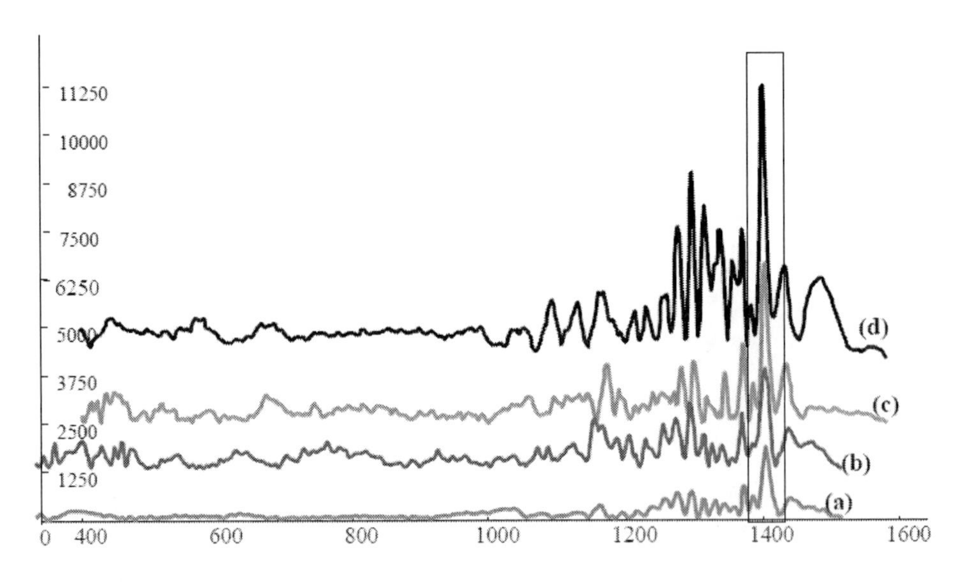

Figure 8. The Raman spectra under VH configureation at zero angle between fiber axis and polarization plane, (a) 22.5 m/min (b) 49.6 m/min (c) 59.5 m/min and (d) 67.7 m/min.

Table 4. The intensity ratios derived from Figure 7 and Figure 8. Based on shown values of enhanced peak at 1394cm[4].

Sample	I_{VV0}	I_{VV90}	I_{VH0}	I_{VV90}/I_{VH0}	I_{VV0}/I_{VH0}	I_{VV0}/I_{VV90}
A	4400	1900	1600	1.1875	2.75	2.31
B	3600	1500	2400	0.625	1.5	2.40
C	4000	1500	4277	0.35071	0.93523	2.67
D	3800	2200	6350	0.34646	0.59843	1.73

The mathematic formulation of Raman intensity in VV and VH modes are represented by the following expressions [11, 12]:

$$I^{VV}(\psi)\alpha\left(\cos^4\psi - \frac{6}{7}\cos^2\psi + \frac{3}{35}\right)\langle P_4(\cos\theta)\rangle + \left(\frac{6}{7}\cos^2\psi - \frac{2}{7}\right)\langle P_2(\cos\theta)\rangle + \frac{1}{5} \qquad \text{Equation 3.1}$$

$$I^{VH}(\psi)\alpha\left(-\cos^4\psi + \cos^2\psi - \frac{4}{35}\right)\langle P_4(\cos\theta)\rangle + \frac{1}{21}\langle P_2(\cos\theta)\rangle + \frac{1}{15} \qquad \text{Equation 3.2}$$

The orientation order parameters of <P2 (cosθ)> and <P4 (cosθ)> which are, respectively, the average values of P2 (cosθ) and P4 (cosθ) for the SWNTs bulk product. The Pi(cosθ) is the Legendre polynomial of degree i which is defined as P2(cosθ)= (3 cos 2 θ-1)/2 and P4(cosθ)= (35 cos 4 θ-30 cos2 θ+3)/8 for the second and fourth degree, respectively. More specifically the <P2(cosθ)> is known as the Herman's orientation factor (f) which varies between values of 1 and 0 corresponding, respectively, to nanotubes fully oriented in the fiber direction and randomly distributed [11].

$$f = \frac{3\left\langle \cos^2\theta \right\rangle - 1}{2} \qquad \text{Equation 3.3}$$

The orientation factors can be determined by solving of following simultaneous algebraic equations given in 3.4 and 3.5. These equations are obtained from equations 3.1 and 3.2 by dividing of both side of these equations and substitution of angles of 0 and 90 degrees for ψ.

$$\frac{I_{G,RBM}^{VV}(\psi=0)}{I_{G,RBM}^{VH}(\psi=0)} = -\frac{24\langle P_4(\cos\theta)\rangle + 60\langle P_2(\cos\theta)\rangle + 21}{12\langle P_4(\cos\theta)\rangle - 5\langle P_2(\cos\theta)\rangle - 7} \quad \text{Equation 03.4}$$

$$\frac{I_{G,RBM}^{VV}(\psi=90)}{I_{G,RBM}^{VH}(\psi=0)} = \frac{-9\langle P_4(\cos\theta)\rangle + 30\langle P_2(\cos\theta)\rangle - 21}{12\langle P_4(\cos\theta)\rangle - 5\langle P_2(\cos\theta)\rangle - 7} \quad \text{Equation 3.5}$$

The left hand side terms of equations of 3.4 and 3.5 are the depolarization ratios that can be experimentally determined. As it can be seen only the intensity at 0° and 90 ° are required to determine the <P2 (cosθ)> and <P4 (cosθ)> for a uniaxially oriented nanofibers. Results calculated from equation 3.4 and 3.5 for Herman orientation factor for different sample shows a range between 0.20 and 0.25 at take up speeds of 67.7 m/min and 59.5 m/min, respectively (Fig 9).

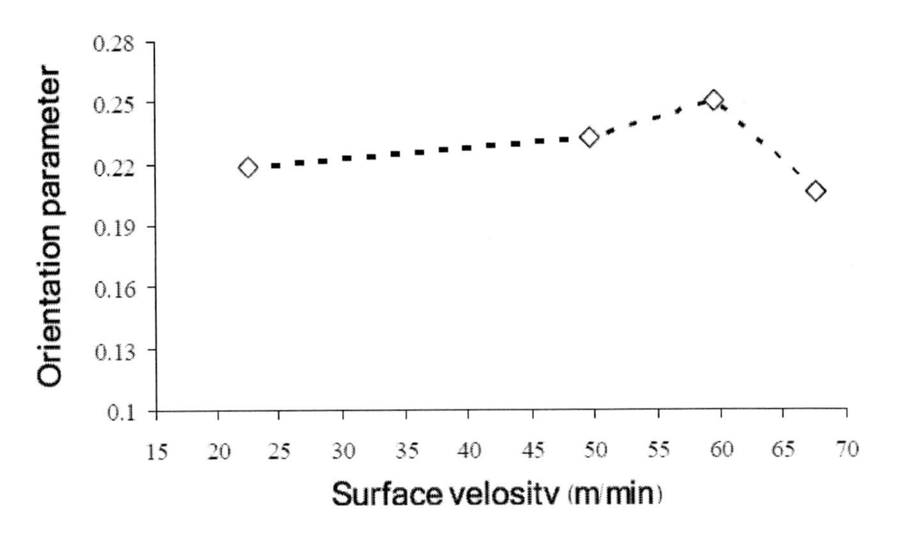

Figure 9. Orientation parameter versus take up speed of rotating drum determined by Raman Spectroscopy.

The maximum chain orientation parameter yield at speed of 59.5 m/min and further increase of it causes loss of molecular orientation which corresponding to other studies [3]. It appears that applying high draw ratio at short time has no significant effect on molecular orientation. Comparing this results with orientation factor of 0.66 and 0.52, which have been observed for commercial wet-spun acrylic fibers and melt-spun acrylic fibers, respectively [13], it can be stated that electrospun PAN nanofibers have lower molecular orientation than commercial fibers.

Conclusions

In this work, electrospinning was carried out with novel and modified method for production of Polyacrylonitrile (PAN) nanofibers using two needles were placed from opposite directions and a rotating collector. The results showed that collected nanofibers by this method have satisfactory mechanical properties as a nanofiber. Raman spectroscopy was utilized as a new technique for analysis of molecular orientation. It was reasonably exhibited that molecular chains orientation had important and considerable effect on mechanical properties of generated nanofibers. In this study, the best alignment, molecular orientation and tensile strength were acquired for nanofibers prepared at collection speed of 59.5 m/min. Also, it was demonstrated that this take up speed can be good choice for collection and performance thermal treatment on PAN nanofibers.

References

[1] M. Ziabari, V. Mottaghitalab, A. K. Haghi, Simulated image of electrospun nonwoven web of PVA and corresponding nanofiber diameter distribution, *Korean Journal of Chemical engng* , Vol.25, No. 4, pp. 919-922 ,2008.

[2] M. Ziabari, V. Mottaghitalab, A. K. Haghi, Evaluation of electrospun nanofiber pore structure parameters, *Korean Journal of Chemical engng* , Vol.25, No. 4, pp. 923-932 ,2008.

[3] M. Ziabari, V. Mottaghitalab, A. K. Haghi, Distance transform algoritm for measuring nanofiber diameter, *Korean Journal of Chemical engng* , Vol25, No. 4, pp. 905-918 ,2008.

[4] Pan, H., Li, L., Hu, L. and Cui, X., "Continuous aligned polymer fibers produced by a modified electrospinning method", *Polymer*, Vol.47, pp.4901-4904, 2006.

[5] Zussman, E., Chen, X., Ding, W., Calabri, L., Dikin, D.A., Quintana, J. P. and Ruoff, R.S., "Mechanical and structural characterization of electrospun PAN-derived carbon nanofibers", *Carbon*, Vol.43, pp.2175-2185, 2005.

[6] Gu, S.Y., Ren, J. and Wu, Q.L., "Preparation and structures of electrospun PAN nanofibers as a precursor of carbon nanofibers", *Synthetic Metals* , Vol.155, pp.157–161, 2005.

[7] Jalili, R., Morshed, M. and Hosseini Ravandi, S.A., "Fundamental Parameters Affecting Electrospinning of PAN Nanofibers as Uniaxially Aligned Fibers", *Journal of Applied Polymer Science*, Vol.101, pp.4350-4357, 2006.

[8] Causin, V., Marega, C., Schiavone, S. and Marigo, A., "A quantitative differentiation method for acrylic fibers by infrared spectroscopy", *Forensic Science International*, Vol.151, pp.125– 131, 2005.

[9] Mathieu, D. and Grand, A., "Ab initio Hartree-Fock Raman spectra of Polyacrylonitrile", *Polymer,* Vol.39, No.21, pp.5011-5017, 1998.

[10] Huang, Y.S. and Koenig, J.L., "Raman spectra of polyacrylonitrile", *Applied Spectroscopy*, Vol.25, pp.620-622, 1971.

[11] Liu, T. and Kumar, S., "Quantitative characterization of SWNT orientation by polarized Raman spectroscopy", *Chem. Phys. Lett*, Vol.378, pp.257-262, 2003.

[12] Jones, W.J., Thomas, D.K., Thomas, D,W. and Williams, G., "On the determination of order parameters for homogeneous and twisted nematic liquid crystals from Raman spectroscopy", *Journal of Molecular Structure*, Vol.708, pp.145–163, 2004.

[13] DavidsonJ, A., Jung, H.T., HudsonS, D. and Percec, S., "Investigation of molecular orientation in melt-spun high acrylonitrile fibers", *Polymer*, Vol.41, pp.3357–3364, 2000.

[14] Soulis, S. and Simitzis, J., "Thermomechanical behaviour of poly [acrylonitrile-*co*-(methyl acrylate)] fibres oxidatively treated at temperatures up to 180 ◦C", *Polymer International,* Vol. 54, pp.1474–1483, 2005.

[15] M.S.A., Rahaman., A.F., Ismail. and A., Mustafa., "A review of heat treatment on polyacrylonitrile fiber", Polymer Degradation and Stability, Vol.92, pp.1421-1432, 2007.

[16] SEN, K., BAJAJ, P. and SREEKUMAR, T.V., "Thermal Behavior of Drawn Acrylic Fibers", *Journal of Polymer Science:Part B:Polymer Physics*, Vol.41, pp.2949–2958, 2003.

[17] Fennessey, s.f., *continuous carbon nanofibers prepared from electrospun polyacrylonitrile precursor fibers*, Polymer Science and Engineering, University of Massachusetts Amherst, 2006.

Chapter 6

ELECTROSPUN BIODEGRADABLE NANOFIBERS

1. INTRODUCTION

In recent years, the electrospinning process has gained much attention because it is an effective method to manufacture ultrafine fibers or fibrous structures of many polymers with diameter in the range from several micrometers down to tens of nanometers [1]. In the electrospinning process, a high voltage is used to create an electrically charged jet of a polymer solution or a molten polymer. This jet is collected on a target as a non-woven fabric. The jet typically develops a bending instability and then solidifies to form fibers, which measures in the range of nanometers to 1 mm. Because these nanofibers have some useful properties such as high specific surface area and high porosity, they can be used as filters, wound dressings, tissue engineering scaffolds, etc.

Recently, the protein-based materials have been interested in biomedical and biotechnological fields. The silk polymer, a representative fibrous protein, has been investigated as one of promising resources of biotechnology and biomedical materials due to its unique properties [2]. Furthermore, high molecular weight synthetic polypeptides of precisely controlled amino acid composition and sequence have been fabricated using the recombinant tools of molecular biology. These genetic engineered polypeptides have attracted the researcher's attention as a new functional material for biotechnological applications including cellular adhesion promoters, biosensors, and suture materials.

Silks are generally defined as fibrous proteins that are spun into fibers by some Lepidoptera larvae such as silkworms, spiders, scorpions, mites and flies. Silkworm silk has been used as a luxury textile material since 3000 B.C., but it was not until recently that the scientific community realized the tremendous potential of silk as a structural material. While stiff and strong fibers (such as carbon, aramid, glass, etc.) are routinely manufactured nowadays, silk fibers offer a unique combination of strength and ductility which is unrivalled by any other natural or man-made fibers. Silk has excellent Properties such as lightweight (1.3 g/cm^3) and high tensile strength (up to 4.8 GPa as the strongest fiber known in nature). Silk is thermally stable up to 250 °C, allowing processing over a wide range of temperatures[3].The origins of this behavior are to be found in the organization of the silks at the molecular and supramolecular levels, which are able to impart a large load bearing capability together with a damage tolerant response. The analysis of the fracture micro mechanisms in silk may shed light on the relationship between its microstructure and the mechanical properties. The anti-parallel β pleated structure of silk gives rise to its structural properties of combined strength and toughness [3].

B. mori silk consists of two types of proteins, fibroin and sericin. Fibroin is the protein that forms the filaments of silkworm silk [3]. Fibroin filaments made up of bundles of nanofibrils with a bundle diameter of around 100 nm. The nanofibrils are oriented parallel to the axis of the fiber, and are thought to interact strongly with each other [3]. Fibroin contains 46% Glycine, 29% Alanine and 12% Serine. Fibroin is a giant molecule (4700 amino acids) comprising a "crystalline" portion of about two-thirds and an "amorphous" region of about one-third. The crystalline portion comprises about 50 repeats of polypeptide of 59 amino acids whose sequence is known: Gly-Ala-Gly-Ala-Gly-Scr-Gly-Ala-Ala-Gly- (Scr-Gly-Ala-Gly-Ala-Gly)s-Tyr .This repeated unit forms, β - sheet and is responsible for the mechanical properties of the fiber [1].

Silk fibroin (SF) can be prepared in various forms such as gels, powders, fibers, and membranes [9].Number of researchers has investigated silk-based nanofibers as one of the candidate materials for biomedical applications, because it has several distinctive biological properties including good biocompatibility, good oxygen and water vapor permeability, biodegradability, and minimal inflammatory reaction. Several researchers have also studied processing parameters and morphology of electrospun silk nanofibers using Hexafluoroacetone, Hexafluoro-2-propanol and formic acid as solvents. In all these reports nanofibers with circular cross sections have been observed.

In the present paper, effects of electrospinning parameters are studied and nanofibers dimensions and morphology are reported. Morphology of fibers diameter of silk precursor were investigated varying concentration, temperature and applied voltage and observation of ribbon like silk nanofibers were reported. Furthermore, a more systematic understanding of the process conditions was studied and a quantitative basis for the relationships between average fiber diameter and electrospinning parameters was established using response surface methodology (RSM), which will provide some basis for the preparation of silk nanofibers with desired properties (Appendix).

2. Experiment

2.1. Preparation of Regenerated SF Solution

Raw silk fibers (B.mori cocoons were obtained from domestic producer, Abrisham Guilan Co., IRAN) were degummed with 2 gr/L Na_2CO_3 solution and 10 gr/L anionic detergent at 100 ° C for 1 h and then rinsed with warm distilled water. Degummed silk (SF) was dissolved in a ternary solvent system of $CaCl_2/CH_3CH_2OH/H_2O$ (1:2:8 in molar ratio) at 70 ° C for 6 h. After dialysis with cellulose tubular membrane (Bialysis Tubing D9527 Sigma) in H_2O for 3 days, the SF solution was filtered and lyophilized to obtain the regenerated SF sponges.

2.2. Preparation of the Spinning Solution

SF solutions were prepared by dissolving the regenerated SF sponges in 98% formic acid for 30 min. Concentrations of SF solutions for electrospinning was in the range from 8% to 14% by weight.

2.3. Electrospinning

In the electrospinning process, a high electric potential (Gamma High voltage) was applied to a droplet of SF solution at the tip (0.35 mm inner diameter) of a syringe needle, The electrospun nanofibers were collected on a target plate which was placed at a distance of 10 cm from the syringe tip. The syringe tip and the target plate were enclosed in a chamber for adjusting and controlling the temperature. Schematic diagram of the electrospinning apparatus is shown in Figure 1. The processing temperature was adjusted at 25, 50 and 75 °C. A high voltage in the range from 10 kV to 20 kV was applied to the droplet of SF solution.

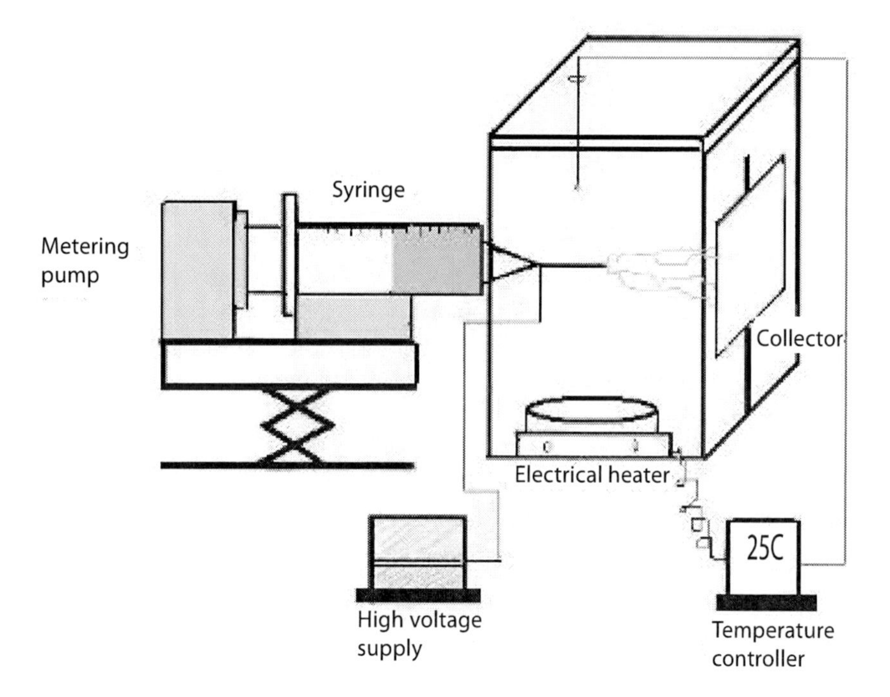

Figure 1. Schematic diagram of electrospinning apparatus.

3. Characterization

Optical microscope (Nikon Microphot-FXA) was used to investigate the macroscopic morphology of electrospun SF fibers. For better resolving power, morphology, surface texture and dimensions of the gold-sputtered electrospun nanofibers were determined using a Philips

XL-30 scanning electron microscope. A measurement of about 100 random fibers was used to determine average fiber diameter and their distribution.

4. Results and Discussion

4.1. Effect of Silk Concentration

One of the most important quantities related with electrospun nanofibers is their diameter. Since nanofibers are resulted from evaporation of polymer jets, the fiber diameters will depend on the jet sizes and the solution concentration. It has been reported that during the traveling of a polymer jet from the syringe tip to the collector, the primary jet may be split into different sizes multiple jets, resulting in different fiber diameters. When no splitting is involved in electrospinning, one of the most important parameters influencing the fiber diameter is Concentration of regenerated silk solution. The jet with a low concentration breaks into droplets readily and a mixture of fibers, bead fibers and droplets as a result of low viscosity is generated. These fibers have an irregular morphology with large variation in size, on the other hand jet with high concentration don't break up but traveled to the grounded target and tend to facilitate the formation of fibers without beads and droplets. In this case, Fibers became more uniform with regular morphology [4-10].

At first, a series of experiments were carried out when the silk concentration was varied from 8 to 14% at the 15KV constant electric field and 25 ° C constant temperature. Below the silk concentration of 8% as well as at low electric filed in the case of 8% solution, droplets were formed instead of fibers. Fig. 2 shows morphology of the obtained fibers from 8% silk solution at 20 KV. The obtained fibers are not uniform. The average fiber diameter is 72 nm and a narrow distribution of fiber diameters is observed. It was found that continues nanofibers were formed above silk concentration of 8% regardless of the applied electric field and electrospinning condition.

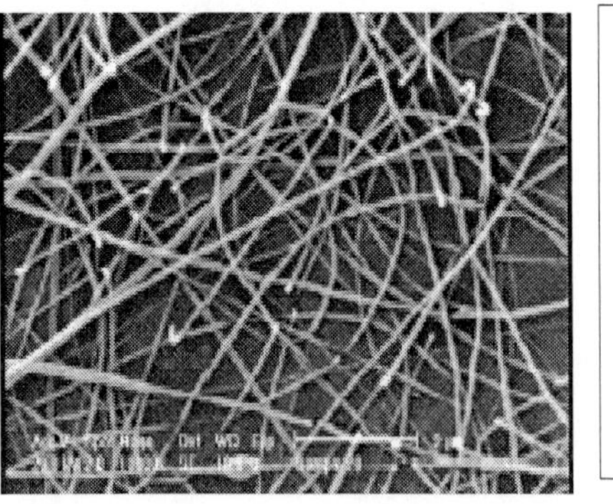

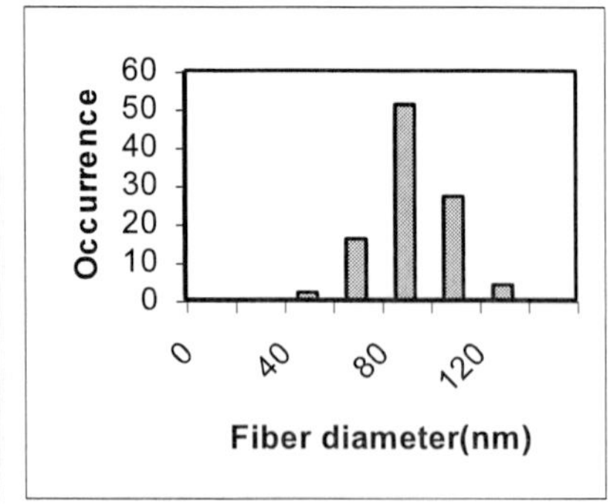

Figure 2. SEM micrograph and fiber distribution of 8wt% of silk at 20 KV and 25 ° C.

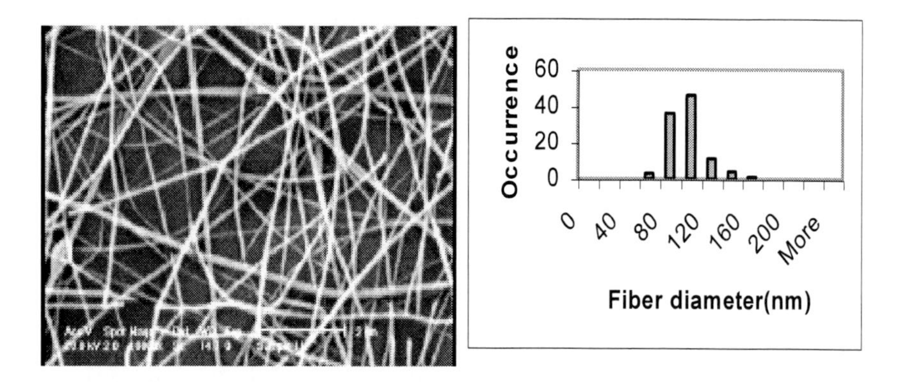

Figure 3.SEM micrograph and fiber distribution of 10wt% of silk at 15 KV and 25 ° C.

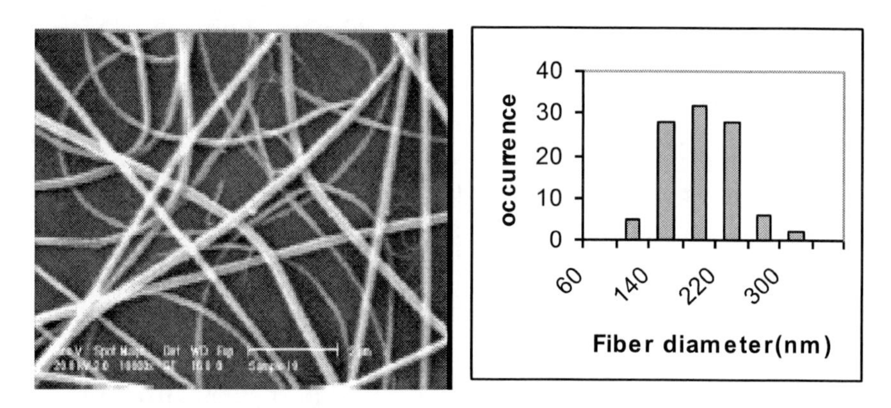

Figure 4.SEM micrograph and fiber distribution of 12wt% of silk at 15 KV and 25 ° C.

In the electrospinning of silk fibroin, when the silk concentration is more than 10%, thin and rod like fibers with diameters range from 60-450 nm were obtained. Figs 3-5 show the SEM micrographs and diameter distribution of the resulted fibers.

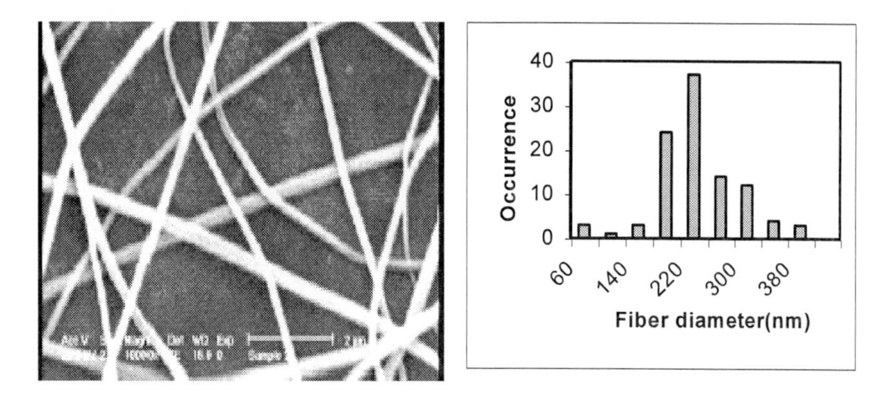

Figure 5.SEM micrograph and fiber distribution of 14wt% of silk at 15 KV and 25 ° C.

There is a significant increase in mean fiber diameter with the increasing of the silk concentration, which shows the important role of silk concentration in fiber formation during electrospinning process. Concentration of the polymer solution reflects the number of entanglements of polymer chains in the solution, thus solution viscosity. Experimental observations in electrospinning confirm that for forming fibers, a minimum polymer concentration is required. Below this critical concentration, application of electric field to a polymer solution results electrospraying and formation of droplets to the instability of the ejected jet. As the polymer concentration increased, a mixture of beads and fibers is formed. Further increase in concentration results in formation of continuous fibers as reported in this paper. It seems that the critical concentration of the silk solution in formic acid for the formation of continuous silk fibers is 10%.

Experimental results in electrospinning showed that with increasing the temperature of electrospinning process, concentration of polymer solution has the same effect on fibers diameter at 25 °C. Figs 6-8 show the SEM micrographs and diameter distribution of the fibers at 50 °C.

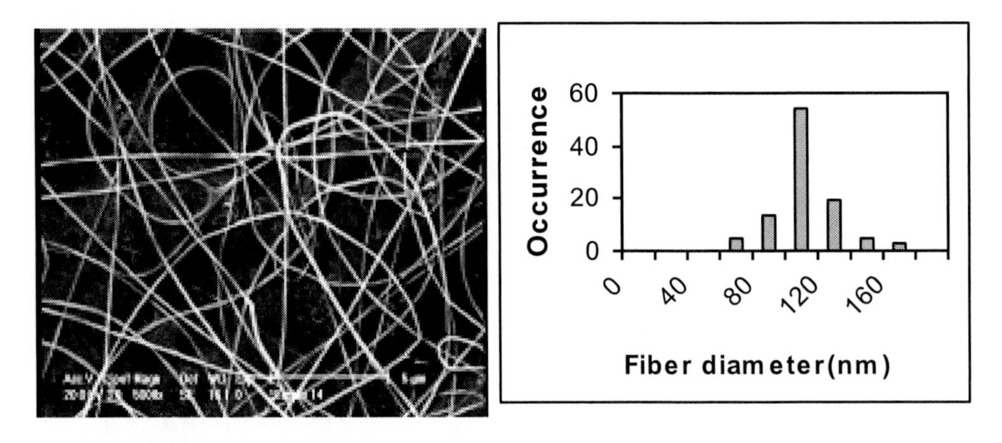

Figure 6.SEM micrograph and fiber distribution of 10wt% of silk at 15 KV and 50 ° C.

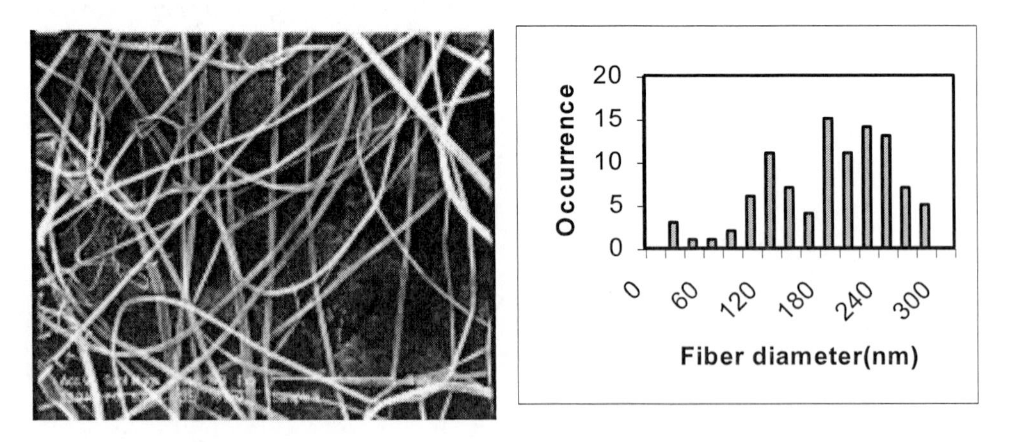

Figure 7.SEM micrograph and fiber distribution of 12wt% of silk at 15 KV and 50 ° C.

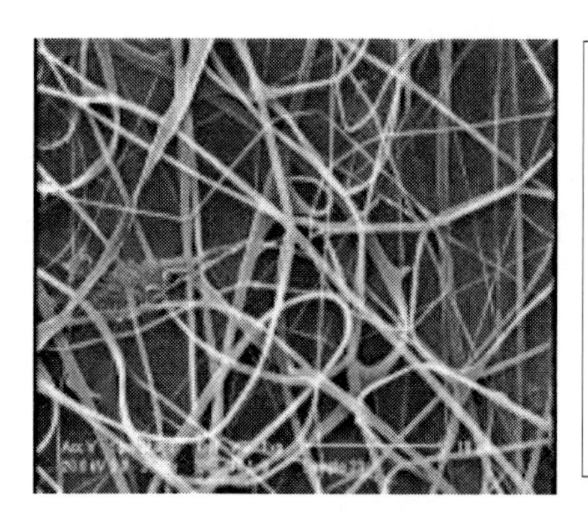

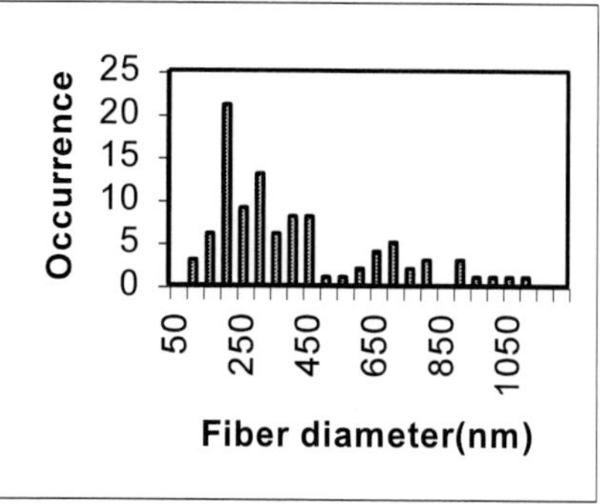

Figure 8.SEM micrograph and fiber distribution of 14wt% of silk at 15 KV and 50 ° C.

When the temperature of the electrospinning process increased, the circular cross section became elliptical and then flat, forming a ribbon with a cross-sectional perimeter nearly the same as the perimeter of the jet. Flat and ribbon like fibers have greater diameter than circular fibers. It seems that at high voltage the primary jet splits into different sizes multiple jets, resulting in different fiber diameters that nearly in a same range of diameter.

At 75 ° C, the concentration and applied voltage have same effect as 50 ° C. Figs 9-11 show the SEM micrographs and diameter distribution of nano fibers at 75 °C.

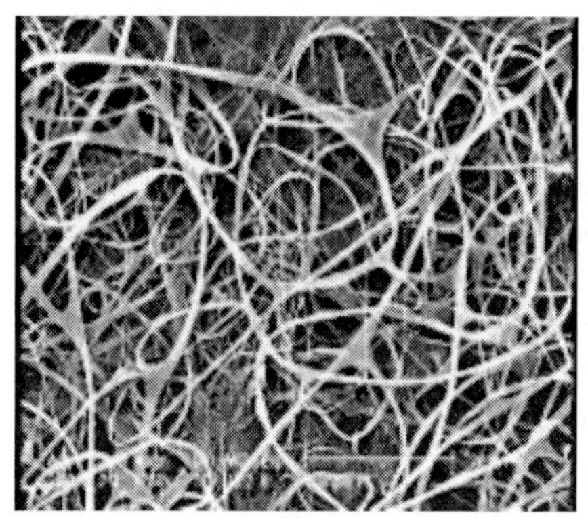

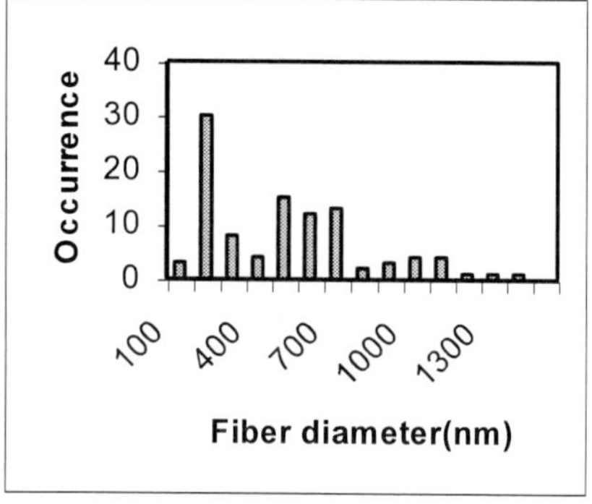

Figure 9.SEM micrograph and fiber distribution of 10wt% of silk at 15 KV and 75 ° C.

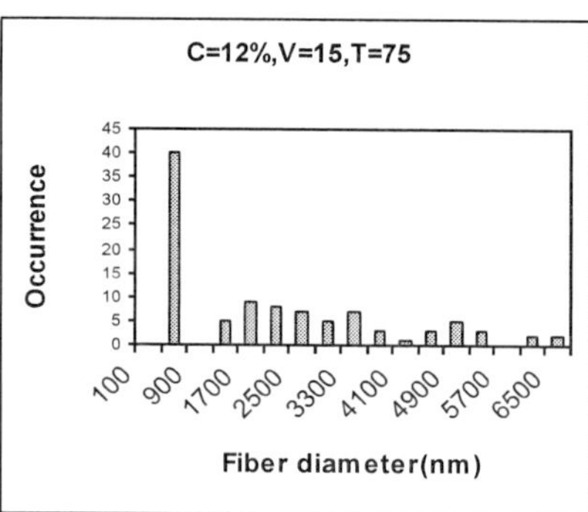

Figure 10.SEM micrograph and fiber distribution of 12wt% of silk at 15 KV and 75 ° C.

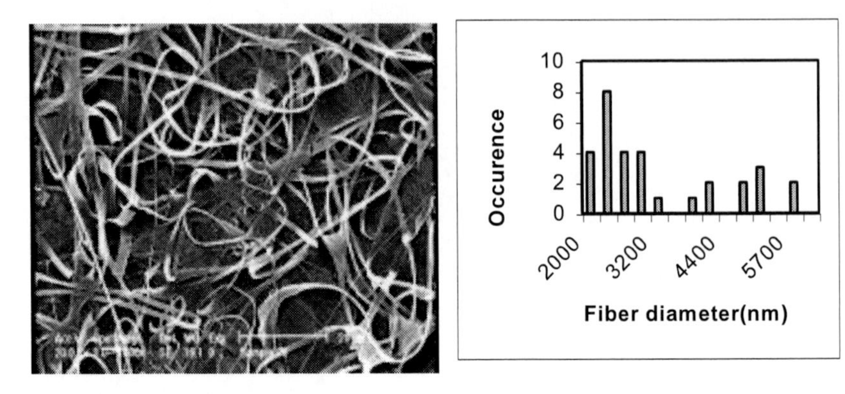

Figure 11.SEM micrograph and fiber distribution of 14wt% of silk at 15 KV and 75 ° C.

Figure 12 shows the relationship between mean fiber diameter and SF concentration at different electrospinning temperature. There is a significant increase in mean fiber diameter with increasing of the silk concentration, which shows the important role of silk concentration in fiber formation during electrospinning process. It is well known that the viscosity of polymer solutions is proportional to concentration and polymer molecular weight. For concentrated polymer solution, concentration of the polymer solution reflects the number of entanglements of polymer chains, thus have considerable effects on the solution viscosity. At fixed polymer molecular weight, the higher polymer concentration resulting higher solution viscosity. The jet from low viscosity liquids breaks up into droplets more readily and few fibers are formed, while at high viscosity, electrospinning is prohibit because of the instability flow causes by the high cohesiveness of the solution. Experimental observations in electrospinning confirm that for fiber formation to occur, a minimum polymer concentration is required. Below this critical concentration, application of electric field to a polymer solution results electro spraying and formation of droplets to the instability of the ejected jet. As the polymer concentration increased, a mixture of beads and fibers is formed. Further increase in concentration results in formation of continuous fibers as reported in this chapter. It seems that the critical concentration of the silk solution in formic acid for the formation of continuous silk fibers is 10% when the applied electric field was in the range of 10 to 20 kV.

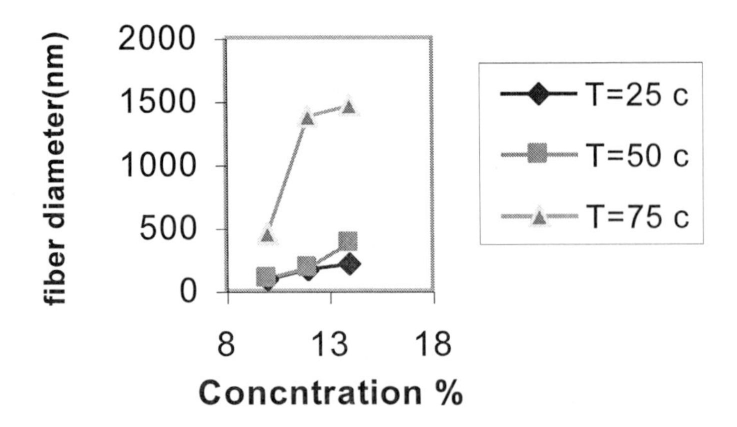

Figure 12. Mean fiber diameter of electrospun silk fibers at 25, 50 and 75 °C at15 kV.

4.2. Effect of Electric Field

It was already reported that the effect of the applied electrospinning voltage is much lower than effect of the solution concentration on the diameter of electrospun fibers. In order to study the effect of the electric field, silk solution with the concentration of 10%, 12%, and 14% were electrospun at 10, 15, and 20 KV at 25 °C. The variation of the mean fiber diameter at different applied voltage for different concentration is shown in fig. 13. This figure and SEM micrographs show that at all electric fields continuous and uniform fibers with different fiber diameter are formed. At a high solution concentration, Effect of applied voltage is nearly significant. It is suggested that, at this temperature, higher applied voltage causes multiple jets formation, which would provide decrees fiber diameter.

As the results of this finding it seems that electric field shows different effects on the nanofibers morphology. This effect depends on the polymer solution concentration and electrospinning conditions.

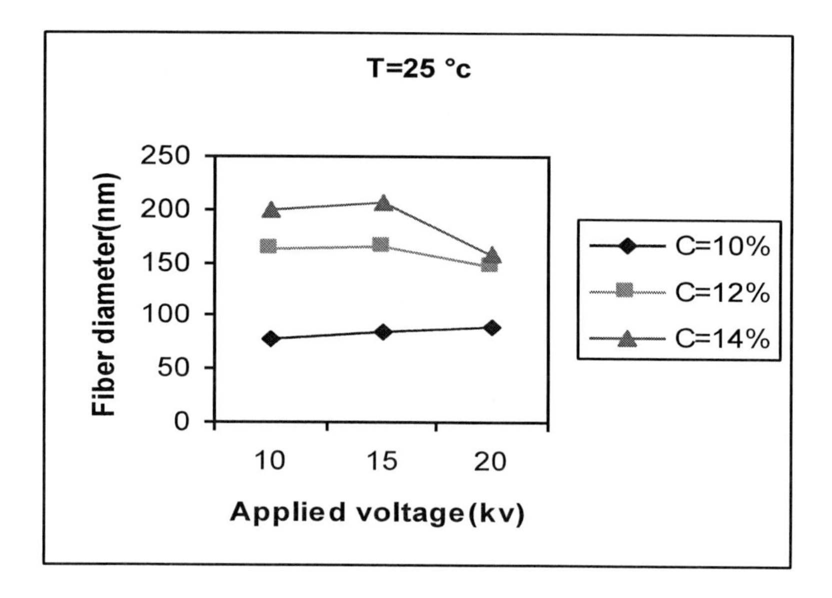

Figure 13.mean fiber diameter of electrospun silk fibers at 10%, 15%, 20% and 25 ° C.

4.3. Effect of Electrospinning Temperature

One of the most important quantities related with electrospun nanofibers is their diameter. Since nanofibers are resulted from evaporation of polymer jets, the fiber diameters will depend on the jet sizes. The elongation of the jet and the evaporation of the solvent both change the shape and the charge per unit area carried by the jet. After the skin is formed, the solvent inside the jet escapes and the atmospheric pressure tends to collapse the tube like jet. The circular cross section becomes elliptical and then flat, forming a ribbon-like structure. In this work we believe that ribbon-like structure in the electrospinning of SF at higher temperature thought to be related with skin formation at the jets. With increasing the electrospinning temperature, solvent evaporation rate increases, which results in the formation of skin at the jet surface. Non- uniform lateral stresses around the fiber due to the uneven evaporation of solvent and/or striking the target make the nanofibers with circular cross-section to collapse into ribbon shape.

Bending of the electrospun ribbons were observed on the SEM micrographs as a result of the electrically driven bending instability or forces that occurred when the ribbon was stopped on the collector. Another problem that may be occurring in the electrospinning of SF at high temperature is the branching of jets. With increasing the temperature of electrospinning process, the balance between the surface tension and electrical forces can shift so that the shape of a jet becomes unstable. Such an unstable jet can reduce its local charge per unit surface area by ejecting a smaller jet from the surface of the primary jet or by splitting apart into two smaller jets. Branched jets, resulting from the ejection of the smaller jet on the surface of the primary jet were observed in electrospun fibers of SF. The axes of the cones from which the secondary jets originated were at an angle near 90° with respect to the axis of the primary jet. Fig. 14 shows the SEM micrographs of flat, ribbon like and branched fibers.

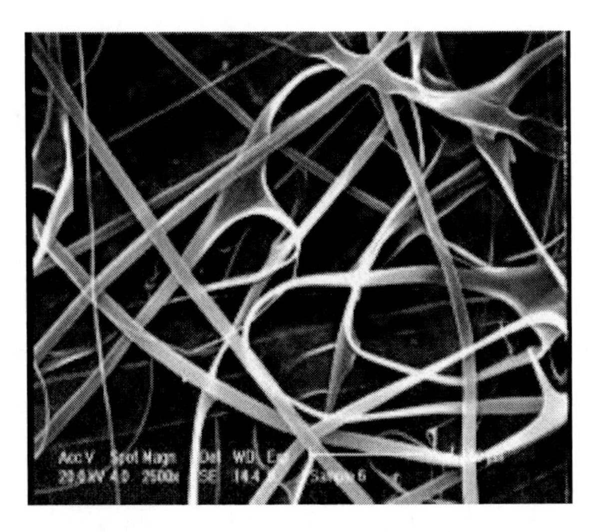

Figure14. SEM micrograph of silk flat, ribbon like and branched nano fibers at high temperature.

In order to study the effect of electrospinning temperature on the morphology and texture of electrospun silk nanofibers, 12% silk solution was electrospun at various temperatures of 25, 50 and 75 °C. Results are shown in Fig. 15. Interestingly, the electrospinning of silk solution showed flat fiber morphology at 50 and 75 °C, whereas circular structure was observed at 25 °C. At 25 °C, the nanofibers with a rounded cross section and a smooth surface were collected on the target. Their diameter showed a size range of approximately 100 to 300 nm with 180 nm being the most frequently occurring. They are within the same range of reported size for electrospun silk nanofibers. With increasing the electrospinning temperature to 50 °C, The morphology of the fibers was slightly changed from circular cross section to ribbon like fibers. Fiber diameter was also increased to a range of approximately 20 to 320 nm with 180 nm the most occurring frequency. At 75 °C, The morphology of the fibers was completely changed to ribbon like structure. Furthermore, fibers dimensions were increased significantly to the range of 500 to 4100 nm with 1100 nm the most occurring frequency.

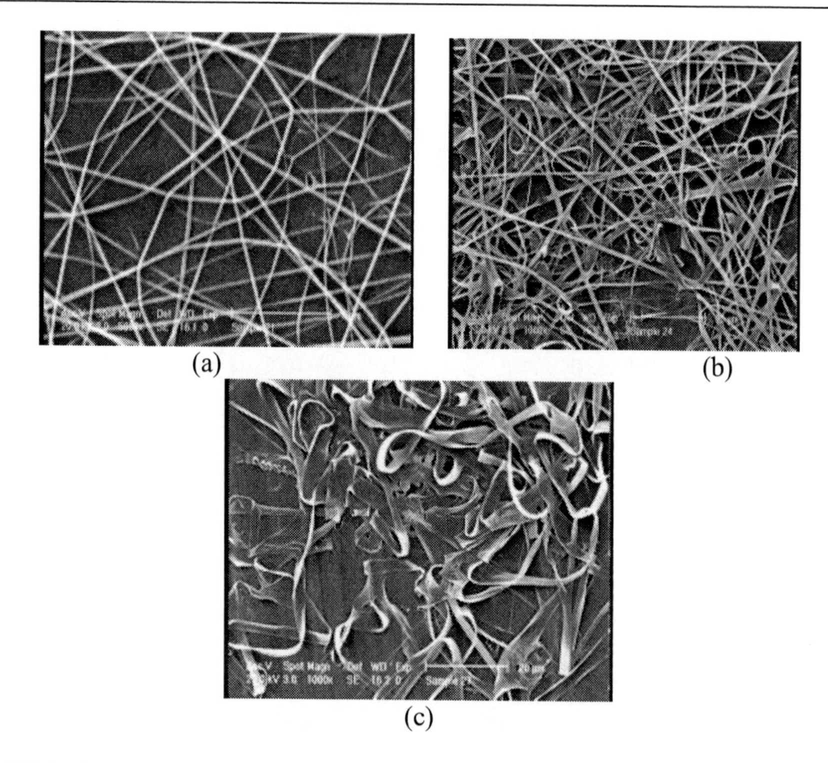

Figure15. SEM micrograph of 12wt% of silk at 20 KV and (a) 25 ° C ,(b) 50 ° C, (c) 75 ° C.

5. EXPERIMENTAL DESIGN

Response surface methodology (RSM) is a collection of mathematical and statistical techniques for empirical model building (Appendix). By careful design of *experiments*, the objective is to optimize a *response* (output variable) which is influenced by several *independent variables* (input variables). An experiment is a series of tests, called *runs*, in which changes are made in the input variables in order to identify the reasons for changes in the output response.

In order to optimize and predict the morphology and average fiber diameter of electrospun silk, design of experiment was employed in the present work. Morphology of fibers and distribution of fiber diameter of silk precursor were investigated varying concentration, temperature and applied voltage. A more systematic understanding of these process conditions was obtained and a quantitative basis for the relationships between average fiber diameter and electrospinning parameters was established using response surface methodology (Appendix), which will provide some basis for the preparation of silk nanofibers.

A central composite design was employed to fit a second-order model for three variables. Silk concentration (X_1), applied voltage (X_2), and temperature (X_3) were three independent variables (factors) considered in the preparation of silk nanofibers, while the fibers diameter were dependent variables (response). The actual and corresponding coded values of three factors (X_1, X_2, and X_3) are given in table 1. The following second-order model in X_1, X_2 and X_3 was fitted using the data in table 1:

$$Y = \beta_0 + \beta_1 x_1 + \beta_2 x_2 + \beta_3 x_3 + \beta_{11} x_1^2 + \beta_{22} x_2^2 + \beta_{33} x_3^2 + \beta_{12} x_1 x_2 + \beta_{13} x_1 x_3 + \beta_{23} x_2 x_3 + \varepsilon$$

Table 1. Central composite design

		Coded values		
X_i	Independent variables	-1	0	1
X_1	Silk concentration (%)	10	12	14
X_2	applied voltage (KV)	10	15	20
X_3	temperature (° C)	25	50	75

The Minitab and Mathlab programs were used for analysis of this second-order model and for response surface plots (Minitab 11, Mathlab 7).

By Regression analysis, values for coefficients for parameters and P-values (a measure of the statistical significance) are calculated. When P-value is less than 0.05, the factor has significant impact on the average fiber diameter. If P-value is greater than 0.05, the factor has no significant impact on average fiber diameter. And R^2_{adj} (represents the proportion of the total variability that has been explained by the regression model) for regression models were obtained (Table 2) and main effect plots on fiber diameter (Fig 16) were obtained and reported.

The fitted second-order equation for average fiber diameter is given by:

$$Y = 391 + 311 X_1 - 164 X_2 + 57 X_3 - 162 X_1^2 + 69 X_2^2 + 391 X_3^2 - 159 X_1X_2 + 315 X_1X_3 - 144 X_2X_3 \quad (1)$$

Where Y = Average fiber diameter

Table 2. Regression Analysis for the three factors (concentration, applied voltage, temperature) and coefficients of the model in coded unit*

Variables	Constant		P-value
	β_0	391.3	0.008
x_1	β_1	310.98	0.00
x_2	β_2	-164.0	0.015
x_3	β_3	57.03	0.00
x^2_1	β_{11}	161.8	0.143
x^2_2	β_{22}	68.8	0.516
x^2_3	β_{33}	390.9	0.002
x_1x_2	β_{12}	-158.77	0.048
x_1x_3	β_{13}	314.59	0.001
x_2x_3	β_{23}	-144.41	0.069
F	P-value	R^2	R^2 (adj)
18.84	0.00	0.907	0.858

* Model: $Y=\beta o+\beta_1 x_{1+}\ \beta_2 x_{2+}\ \beta_3 x_3 + \beta_{11} x^2_1 + \beta_{22} x^2_2 + \beta_{332} x^2_{3\ +}\ \beta_{12} x_1 x2_+\ \beta_{13} x_1 x\ 3_+\ \beta_{13} x_2 x3$ where "y" is average fiber diameter.

From the P-values listed in Table 2, it is obvious that P-value of term X_2 is greater than P-values for terms X_1 and X_3. And other P-values for terms related to applied voltage such as, X_2^2, X_1X_2, X_2X_3 are much greater than significance level of 0.05. That is to say, applied voltage has no much significant impact on average fiber diameter and the interactions between concentration and applied voltage, temperature and applied voltage are not significant, either. But P-values for term related to X_3 and X_1 are less than 0.05. Therefore, temperature and concentration have significant impact on average fiber diameter. Furthermore, R^2_{adj} is 0.858, That is to say, this model explains 86% of the variability in new data.

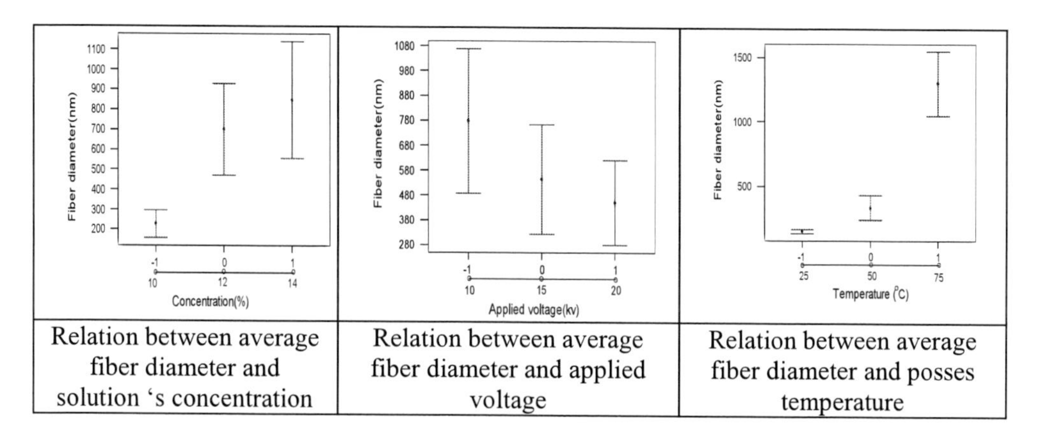

Relation between average fiber diameter and solution ‘s concentration	Relation between average fiber diameter and applied voltage	Relation between average fiber diameter and posses temperature

Figure16. Effect of electrospinning parameters on silk nano fibers diameter.

CONCLUSION

The electrospinning of silk fibroin was processed and the average fiber diameters depend on the electrospinning condition. Morphology of fibers and distribution of diameter were investigated at various concentrations, applied voltages and temperature. The electrospinning temperature and the solution concentration have a significant effect on the morphology of the electrospun silk nanofibers. There effects were explained to be due to the change in the rate of skin formation and the evaporation rate of solvents. To determine the exact mechanism of the conversion of polymer into nanofibers require further theoretical and experimental work.

From the practical view the results of the present work can be condensed. Concentration of regenerated silk solution was the most dominant parameter to produce uniform and continuous fibers. The jet with a low concentration breaks into droplets readily and a mixture of fibers and droplets as a result of low viscosity is generated. On the other hand jets with high concentration do not break up but traveled to the target and tend to facilitate the formation of fibers without beads and droplets. In this case, fibers become more uniform with regular morphology. In the electrospinning of silk fibroin, when the silk concentration is more than 10%, thin and rod like fibers with diameters range from 60-450 nm were obtained. Furthermore, In the electrospinning of silk fibroin, when the process temperature is more than 25 °C, flat, ribbon like and branched fibers with diameters range from 60-7000 nm were obtained.

Two- way analysis of variance was carried out at the significant level of 0.05 to study the impact of concentration, applied voltages and temperature on average fiber diameter. It was concluded that concentration of solution and electrospinning temperature were the most significant factors impacting the diameter of fibers. Applied voltage had no significant impact on average fiber diameter. The average fiber diameter increased with polymer concentration and electrospinning temperature according to proposed relationship under the experimental conditions studied in this chapter.

APPENDIX

Variables which potentially can alter the electrospinning process (Figure A-1) are large. Hence, investigating all of them in the framework of one single research would almost be impossible. However, some of these parameters can be held constant during experimentation. For instance, performing the experiments in a controlled environmental condition, which is concerned in this study, the ambient parameters (i.e. temperature, air pressure, and humidity) are kept unchanged. Solution viscosity is affected by polymer molecular weight, solution concentration, and temperature. For a particular polymer (constant molecular weight) at a fixed temperature, solution concentration would be the only factor influencing the viscosity. In this circumstance, the effect of viscosity could be determined by the solution concentration. Therefore, there would be no need for viscosity to be considered as a separate parameter.

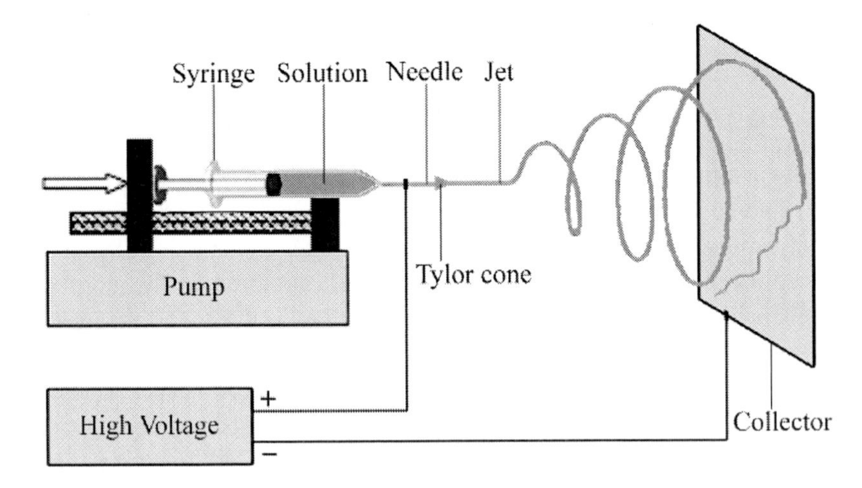

Figure A-1. A typical image of Electrospinning process.

In this regard, solution concentration (*C*), spinning distance (*d*), applied voltage (*V*), and volume flow rate (*Q*) were selected to be the most influential parameters. The next step is to choose the ranges over which these factors are varied. Process knowledge, which is a combination of practical experience and theoretical understanding, is required to fulfill this step. The aim is here to find an appropriate range for each parameter where dry, bead-free, stable, and continuous fibers without breaking up to droplets are obtained. This goal could be achieved by conducting a set of preliminary experiments while having the previous works in mind along with utilizing the reported relationships.

The relationship between intrinsic viscosity ($[\eta]$) and molecular weight (M) is given by the well-known Mark-Houwink-Sakurada equation as follows:

$$[\eta] = KM^{a} \quad \text{(A-1)}$$

where K and a are constants for a particular polymer-solvent pair at a given temperature. Polymer chain entanglements in a solution can be expressed in terms of Berry number (B), which is a dimensionless parameter and defined as the product of intrinsic viscosity and polymer concentration ($B=[\eta]C$). For each molecular weight, there is a lower critical concentration at which the polymer solution cannot be electrospun.

As for determining the appropriate range of applied voltage, referring to previous works, it was observed that the changes of voltage lay between 5 *kV* to 25 *kV* depending on experimental conditions; voltages above 25 *kV* were rarely used. Afterwards, a series of experiments were carried out to obtain the desired voltage domain. At $V<10$ *kV*, the voltage was too low to spin fibers and $10\ kV \leq V < 15\ kV$ resulted in formation of fibers and droplets; in addition, electrospinning was impeded at high concentrations. In this regard, $15\ kV \leq V \leq 25\ kV$ was selected to be the desired domain for applied voltage.

The use of 5 *cm* – 20 *cm* for spinning distance was reported in the literature. Short distances are suitable for highly evaporative solvents whereas it results in wet coagulated fibers for nonvolatile solvents due to insufficient evaporation time. Afterwards, this was proved by experimental observations and $10\ cm \leq d \leq 20\ cm$ was considered as the effective range for spinning distance.

Few researchers have addressed the effect of volume flow rate. Therefore in this case, the attention was focused on experimental observations. At $Q<0.2$ *ml/h*, in most cases especially at high polymer concentrations, the fiber formation was hindered due to insufficient supply of solution to the tip of the syringe needle. Whereas, excessive feed of solution at $Q>0.4$ *ml/h* incurred formation of droplets along with fibers. As a result, $0.2\ ml/h \leq Q \leq 0.4\ ml/h$ was chosen as the favorable range of flow rate in this study.

Consider a process in which several factors affect a response of the system. In this case, a conventional strategy of experimentation, which is extensively used in practice, is the *one-factor-at-a-time* approach. The major disadvantage of this approach is its failure to consider any possible interaction between the factors, say the failure of one factor to produce the same effect on the response at different levels of another factor. For instance, suppose that two factors A and B affect a response. At one level of A, increasing B causes the response to increase, while at the other level of A, the effect of B totally reverses and the response decreases with increasing B. As interactions exist between electrospinning parameters, this approach may not be an appropriate choice for the case of the present work. The correct strategy to deal with several factors is to use a full factorial design. In this method, factors are all varied together; therefore all possible combinations of the levels of the factors are investigated. This approach is very efficient, makes the most use of the experimental data and takes into account the interactions between factors.

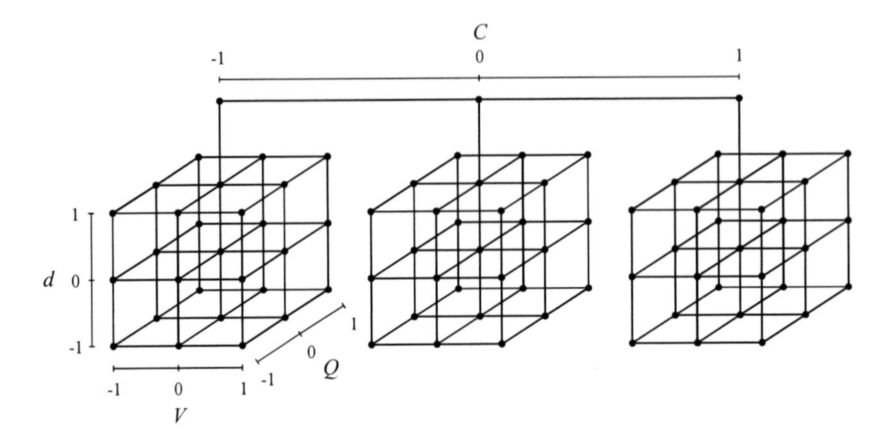

Figure A-2. 3^4 full factorial experimental design used in this study.

It is trivial that in order to draw a line at least two points and for a quadratic curve at least three points are required. Hence, three levels were selected for each parameter in this study so that it would be possible to use quadratic models. These levels were chosen equally spaced. A full factorial experimental design with four factors (solution concentration, spinning distance, applied voltage, and flow rate) each at three levels (3^4 design) were employed resulting in 81 treatment combinations. This design is shown in Figure A-2.

-1, 0, and 1 are coded variables corresponding to low, intermediate and high levels of each factor respectively. The coded variables (x_j) were calculated using Equation (2) from natural variables (ξ_i). The indices 1 to 4 represent solution concentration, spinning distance, applied voltage, and flow rate respectively. In addition to experimental data, 15 treatments inside the design space were selected as test data and used for evaluation of the models. The natural and coded variables for experimental data (numbers 1-81) as well as test data (numbers 82-96) are listed in Table 8 in Appendix.

$$x_j = \frac{\xi_j - [\xi_{hj} + \xi_{lj}]/2}{[\xi_{hj} - \xi_{lj}]/2} \quad \text{(A-2)}$$

The mechanism of some scientific phenomena has been well understood and models depicting the physical behavior of the system have been drawn in the form of mathematical relationships. However, there are numerous processes at the moment which have not been sufficiently been understood to permit the theoretical approach. Response surface methodology (RSM) is a combination of mathematical and statistical techniques useful for empirical modeling and analysis of such systems. The application of RSM is in situations where several input variables are potentially influence some performance measure or quality characteristic of the process – often called responses. The relationship between the response (y) and k input variables ($\xi_1,\xi_2,...,\xi_k$) could be expressed in terms of mathematical notations as follows:

$$y = f(\xi_1, \xi_2, ..., \xi_k) \quad \text{(A-3)}$$

where the true response function f is unknown. It is often convenient to use coded variables (x_1,x_2,..,x_k) instead of natural (input) variables. The response function will then be:

$$y = f(x_1, x_2, ..., x_k) \quad \text{(A-4)}$$

Since the form of true response function f is unknown, it must be approximated. Therefore, the successful use of RSM is critically dependent upon the choice of appropriate function to approximate f. Low-order polynomials are widely used as approximating functions. First order (linear) models are unable to capture the interaction between parameters which is a form of curvature in the true response function. Second order (quadratic) models will be likely to perform well in these circumstances. In general, the quadratic model is in the form of:

$$y = \beta_0 + \sum_{j=1}^{k} \beta_j x_j + \sum_{j=1}^{k} \beta_{jj} x_j^2 + \sum_{i<j} \sum_{j=2}^{k} \beta_{ij} x_i x_j + \varepsilon \quad \text{(A-5)}$$

where ε is the error term in the model. The use of polynomials of higher order is also possible but infrequent. The βs are a set of unknown coefficients needed to be estimated. In order to do that, the first step is to make some observations on the system being studied. The model in Equation (5) may now be written in matrix notations as:

$$\mathbf{y} = \mathbf{X}\boldsymbol{\beta} + \boldsymbol{\varepsilon} \quad \text{(A-6)}$$

where $\mathbf{y}$ is the vector of observations, $\mathbf{X}$ is the matrix of levels of the variables, $\boldsymbol{\beta}$ is the vector of unknown coefficients, and $\boldsymbol{\varepsilon}$ is the vector of random errors. Afterwards, method of least squares, which minimizes the sum of squares of errors, is employed to find the estimators of the coefficients ($\hat{\boldsymbol{\beta}}$) through:

$$\hat{\boldsymbol{\beta}} = (\mathbf{X}'\mathbf{X})^{-1}\mathbf{X}'\mathbf{y} \quad \text{(A-7)}$$

The fitted model will then be written as:

$$\hat{\mathbf{y}} = \mathbf{X}\hat{\boldsymbol{\beta}} \quad \text{(A-8)}$$

Finally, response surfaces or contour plots are depicted to help visualize the relationship between the response and the variables and see the influence of the parameters. As you might notice, there is a close connection between RSM and linear regression analysis.

After the unknown coefficients (βs) were estimated by least squares method, the quadratic models for the mean fiber diameter (MFD) and standard deviation of fiber diameter (StdFD) in terms of coded variables are written as:

$$\begin{aligned}\text{MFD} = {} & 282.031 + 34.953x_1 + 5.622x_2 - 2.113x_3 + 9.013x_4 \\ & - 11.613x_1^2 - 4.304x_2^2 - 15.500x_3^2 \\ & - 0.414x_4^2 + 12.517x_1x_2 + 4.020x_1x_3 - 0.162x_1x_4 + 20.643x_2x_3 + 0.741x_2x_4 + 0.877x_3x_4\end{aligned} \quad \text{(A-9)}$$

$$\begin{aligned}\text{StdFD} = {} & 36.1574 + 4.5788x_1 - 1.5536x_2 + 6.4012x_3 + 1.1531x_4 \\ & - 2.2937x_1^2 - 0.1115x_2^2 - 1.1891x_3^2 + 3.0980x_4^2 \\ & - 0.2088x_1x_2 + 1.0010x_1x_3 + 2.7978x_1x_4 + 0.1649x_2x_3 - 2.4876x_2x_4 + 1.5182x_3x_4\end{aligned} \quad \text{(A-10)}$$

In the next step, a couple of very important hypothesis-testing procedures were carried out to measure the usefulness of the models presented here. First, the test for significance of the model was performed to determine whether there is a subset of variables which contributes significantly in representing the response variations. The appropriate hypotheses are:

$$\begin{aligned}& H_0 : \beta_1 = \beta_2 = \cdots = \beta_k \\ & H_1 : \beta_j \neq 0 \quad \text{for at least one } j\end{aligned} \quad \text{(A-11)}$$

The F statistics (the result of dividing the factor mean square by the error mean square) of this test along with the p-values (a measure of statistical significance, the smallest level of significance for which the null hypothesis is rejected) for both models are shown in Table 1.

Table A-1. Summary of the results from statistical analysis of the models

	F	p-value	R^2	R^2_{adj}	R^2_{pred}
MFD	106.02	0.000	95.74%	94.84%	93.48%
StdFD	42.05	0.000	89.92%	87.78%	84.83%

The p-values of the models are very small (almost zero), therefore it could be concluded that the null hypothesis is rejected in both cases suggesting that there are some significant terms in each model. There are also included in Table 1, the values of R^2, R^2_{adj}, and R^2_{pred}. R^2 is a measure for the amount of response variation which is explained by variables and will always increase when a new term is added to the model regardless of whether the inclusion of the additional term is statistically significant or not. R^2_{adj} is the adjusted form of R^2 for the number of terms in the model; therefore it will increase only if the new terms improve the model and decreases if unnecessary terms are added. R^2_{pred} implies how well the model predicts the response for new observations, whereas R^2 and R^2_{adj} indicate how well the model fits the experimental data. The R^2 values demonstrate that 95.74% of MFD and 89.92% of StdFD are explained by the variables. The R^2_{adj} values are 94.84% and 87.78% for MFD and StdFD respectively, which account for the number of terms in the models. Both R^2 and R^2_{adj} values indicate that the models fit the data very well. The slight difference between the values of R^2 and R^2_{adj} suggests that there might be some insignificant terms in the models. Since the R^2_{pred} values are so close to the values of R^2 and R^2_{adj}, models does not appear to be overfit and have very good predictive ability.

The second testing hypothesis is evaluation of individual coefficients, which would be useful for determination of variables in the models. The hypotheses for testing of the significance of any individual coefficient are:

$$H_0 : \beta_j = 0$$
$$H_1 : \beta_j \neq 0 \qquad \text{(A-12)}$$

Table A-2. The test on individual coefficients for the model of mean fiber diameter (MFD)

Term (coded)	Coeff.	T	p-value
Constant	282.031	102.565	0.000
C	34.953	31.136	0.000
d	5.622	5.008	0.000
V	-2.113	-1.882	0.064
Q	9.013	8.028	0.000
C^2	-11.613	-5.973	0.000
d^2	-4.304	-2.214	0.030
V^2	-15.500	-7.972	0.000
Q^2	-0.414	-0.213	0.832
Cd	12.517	9.104	0.000
CV	4.020	2.924	0.005
CQ	-0.162	-0.118	0.906
dV	20.643	15.015	0.000
dQ	0.741	0.539	0.592
VQ	0.877	0.638	0.526

The model might be more efficient with inclusion or perhaps exclusion of one or more variables. Therefore the value of each term in the model is evaluated using this test, and then eliminating the statistically insignificant terms, more efficient models could be obtained. The results of this test for the models of MFD and StdFD are summarized in Table 2 and Table 3 respectively. T statistic in these tables is a measure of the difference between an observed statistic and its hypothesized population value in units of standard error.

As depicted, the terms related to Q^2, CQ, dQ, and VQ in the model of MFD and related to d^2, Cd, and dV in the model of StdFD have very high p-values, therefore they do not contribute significantly in representing the variation of the corresponding response. Eliminating these terms will enhance the efficiency of the models. The new models are then given by recalculating the unknown coefficients in terms of coded variables in equations (A-13) and(14), and in terms of natural (uncoded) variables in equations(A-15), (A-16).

Table A-3. The test on individual coefficients for the model of standard deviation of fiber diameter (StdFD)

Term (coded)	Coef	T	p-value
Constant	36.1574	39.381	0.000
C	4.5788	12.216	0.000
D	-1.5536	-4.145	0.000
V	6.4012	17.078	0.000
Q	1.1531	3.076	0.003
C^2	-2.2937	-3.533	0.001
d^2	-0.1115	-0.172	0.864
V^2	-1.1891	-1.832	0.072
Q^2	3.0980	4.772	0.000
Cd	-0.2088	-0.455	0.651
CV	1.0010	2.180	0.033
CQ	2.7978	6.095	0.000
dV	0.1649	0.359	0.721
dQ	-2.4876	-5.419	0.000
VQ	1.5182	3.307	0.002

$$\begin{aligned} MFD = {} & 281.755 + 34.953x_1 + 5.622x_2 - 2.113x_3 + 9.013x_4 \\ & - 11.613x_1^2 - 4.304x_2^2 - 15.500x_3^2 \\ & + 12.517x_1x_2 + 4.020x_1x_3 + 20.643x_2x_3 \end{aligned} \quad \text{(A-13)}$$

$$\begin{aligned} StdFD = {} & 36.083 + 4.579x_1 - 1.554x_2 + 6.401x_3 + 1.153x_4 \\ & - 2.294x_1^2 - 1.189x_3^2 + 3.098x_4^2 \\ & + 1.001x_1x_3 + 2.798x_1x_4 - 2.488x_2x_4 + 1.518x_3x_4 \end{aligned} \quad \text{(A-14)}$$

$$\begin{aligned} MFD = {} & 10.3345 + 48.7288\,C - 22.7420\,d + 7.9713\,V + 90.1250\,Q \\ & - 2.9033\,C^2 - 0.1722\,d^2 - 0.6120\,V^2 \\ & + 1.2517\,Cd + 0.4020\,CV + 0.8257\,dV \end{aligned} \quad \text{(A-15)}$$

$$\begin{aligned} StdFD = {} & -1.8823 + 7.5590\,C + 1.1818\,d + 1.2709\,V - 300.3410\,Q \\ & - 0.5734\,C^2 - 0.0476\,V^2 + 309.7999\,Q^2 \\ & + 0.1001\,CV + 13.9892\,CQ - 4.9752\,dQ + 3.0364\,VQ \end{aligned} \quad \text{(A-16)}$$

Table A-4. Summary of the results from statistical analysis of the models after eliminating the insignificant terms

	F	p-value	R^2	R^2_{adj}	R^2_{pred}
MFD	155.56	0.000	95.69%	95.08%	94.18%
StdFD	55.61	0.000	89.86%	88.25%	86.02%

The results of the test for significance as well as R^2, R^2_{adj}, and R^2_{pred} for the new models are given in Table 4. It is obvious that the p-values for the new models are close to zero indicating the existence of some significant terms in each model. Comparing the results of this table with Table 1, the *F* statistic increased for the new models, indicating the improvement of the models after eliminating the insignificant terms. Despite the slight decrease in R^2, the values of R^2_{adj}, and R^2_{pred} increased substantially for the new models. As it was mentioned earlier in the paper, R^2 will always increase with the number of terms in the model. Therefore, the smaller R^2 values were expected for the new models, due to the fewer terms. However, this does not necessarily suggest that the pervious models were more efficient. Looking at the tables, R^2_{adj}, which provides a more useful tool for comparing the explanatory power of models with different number of terms, increased after eliminating the unnecessary variables. Hence, the new models have the ability to better explain the experimental data. Due to higher R^2_{pred}, the new models also have higher prediction ability. In other words, eliminating the insignificant terms results in simpler models which not only present the experimental data in superior form, but also are more powerful in predicting new conditions.

The test for individual coefficients was performed again for the new models. The results of this test are summarized in Table 5 and Table 6. This time, as it was anticipated, no terms had higher *p*-value than expected, which need to be eliminated. Here is another advantage of removing unimportant terms. The values of *T* statistic increased for the terms already in the models implying that their effects on the response became stronger.

Table A-5. The test on individual coefficients for the model of mean fiber diameter (MFD) after eliminating the insignificant terms

Term (coded)	Coeff.	T	*p*-value
Constant	281.755	118.973	0.000
C	34.953	31.884	0.000
d	5.622	5.128	0.000
V	-2.113	-1.927	0.058
Q	9.013	8.221	0.000
C^2	-11.613	-6.116	0.000
d^2	-4.304	-2.267	0.026
V^2	-15.500	-8.163	0.000
Cd	12.517	9.323	0.000
CV	4.020	2.994	0.004
dV	20.643	15.375	0.000

Table A-6. The test on individual coefficients for the model of standard deviation of fiber diameter (StdFD) after eliminating the insignificant terms

Term (coded)	Coef	T	*p*-value
Constant	36.083	45.438	0.000
C	4.579	12.456	0.000
d	-1.554	-4.226	0.000
V	6.401	17.413	0.000
Q	1.153	3.137	0.003
C^2	-2.294	-3.602	0.001
V^2	-1.189	-1.868	0.066
Q^2	3.098	4.866	0.000
CV	1.001	2.223	0.029
CQ	2.798	6.214	0.000
dQ	-2.488	-5.525	0.000
VQ	1.518	3.372	0.001

After developing the relationship between parameters, the test data were used to investigate the prediction ability of the models. Root mean square errors (RMSE) between the calculated responses (C_i) and real responses (R_i) were determined using equation **Error! Reference source not found.** for experimental data as well as test data for the sake of evaluation of both MFD and StdFD models.

$$\mathrm{RMSE} = \sqrt{\frac{\sum_{i=1}^{n}(C_i - R_i)^2}{n}} \quad \text{(A-17)}$$

REFERENCES

[1] M. Ziabari, V. Mottaghitalab, A. K. Haghi, Simulated image of electrospun nonwoven web of PVA and corresponding nanofiber diameter distribution, *Korean Journal of Chemical engng ,* Vol.25, No. 4, pp. 919-922 ,2008.

[2] M. Ziabari, V. Mottaghitalab, A. K. Haghi, Evaluation of electrospun nanofiber pore structure parameters, *Korean Journal of Chemical engng ,* Vol.25, No. 4, pp. 923-932 ,2008.

[3] M. Ziabari, V. Mottaghitalab, A. K. Haghi, Distance transform algoritm for measuring nanofiber diameter, *Korean Journal of Chemical engng* , Vol25, No. 4, pp. 905-918 ,2008.

[4] D. Li and Y. Xia, Electrospinning of Nanofibers: Reinventing the Wheel?, *Advanced Materials*, vol. 16, no. 14, pp. 1151-1170, 2004.

[5] R. Derch, A. Greiner and J.H. Wendorff, Polymer Nanofibers Prepared by Electrospinning, In: J. A. Schwarz, C. I. Contescu and K. Putyera, *Dekker Encyclopedia of Nanoscience and Nanotechnology,* CRC, New York, 2004.

[6] A.K. Haghi and M. Akbari, Trends in Electrospinning of Natural Nanofibers. *Physica Status Solidi* (a), vol. 204, pp. 1830-1834, 2007.

[7] P.W. Gibson, H.L. Schreuder-Gibson and D. Rivin, Electrospun Fiber Mats: Transport Properties, *AIChE Journal*, vol. 45, no. 1, pp. 190-195, 1999.

[8] Z.M. Huang, Y.Z. Zhang, M. Kotaki and S. Ramakrishna, A Review on Polymer Nanofibers by Electrospinning and Their Applications in Nanocomposites, *Composites Science and Technology*, vol. 63, pp. 2223-2253, 2003.

[9] M. Li, M.J. Mondrinos, M.R. Gandhi, F.K. Ko, A.S. Weiss and P.I. Lelkes, Electrospun Protein Fibers as Matrices for Tissue Engineering, *Biomaterials*, vol. 26, pp. 5999-6008, 2005.

[10] E.D. Boland, B.D. Coleman, C.P. Barnes, D.G. Simpson, G.E. Wnek and G.L. Bowlin, Electrospinning Polydioxanone for Biomedical Applications, *Acta Biomaterialia*, vol. 1, pp. 115-123, 2005.

Chapter 7

THE ELECTROLESS PLATING ON NANOFIBERS

INTRODUCTION

Because of the high conductivity of copper, electroless copper plating is currently used to manufacture conductive fabrics with high shielding effectiveness (SE). It can be performed at any step of the textile production, such as yarn, stock, fabric or clothing [1].

Electroless copper plating as a non- electrolytic method of deposition from solution on fabrics has been studied by some researchers [1-9]. The early reported copper electroless deposition method uses a catalytic redox reaction between metal ions and dissolved reduction agent of formaldhyde at high temperature and alkaline medium [1-2]. Despite of technique advantages, such as low cost, excellent conductivity, easy formation of a continuous and uniform coating, experimental safety risks appears through formation of hazardous gasous product during plating process specially for industruial scale.

Further research has been conducted to substitute formaldhdyde with other reducing agents coupled with oxidation accelerator such as sodium hypophosphite and nickel sulphate[4-8]. Incorporation of Nickel and Phosphhrus particles providing good potential for creation of fabrics with a metallic appearance and good handling characteristics. These properies are practically viable If plating process followed by finishing process in optmizied pH and in presence of ferrocyanide.Revealing the performance of electroless plating of Cu-Ni-P alloy on cotton fabrics is an essential research area in textile finishing processing and for technological design[4-9].

The main aim of this paper is to explore the possibility of applying electroless plating of Cu-Ni-P alloy onto cotton fabric to obtain highest level of conductivity, washing and abrasion fastness, room condition durability and EMI shielding effectivness. The fabrication and properties of Cu-Ni-P alloy plated cotton fabric are investigated in accordance with standard testing methods.

EXPERIMENTAL

Cotton fabrics (53×48 count/cm^2, 140 g/m^2, taffeta fabric) were used as substrate. The surface area of each specimen is 100 cm^2 .The electroless copper plating process was conducted by multistep processes: pre-cleaning, sensitisation, activation, electroless Cu-Ni-P alloy deposition and post-treatment.

The fabric specimens (10cm × 10 cm) were cleaned with non-ionic detergent (0.5g/l) and $NaHCO_3$ (0.5g/l) solution for 10 minutes prior to use. The samples then were rinsed in distilled water. Surface sensitization was conducted by immersion of the samples into an aqueous solution containing $SnCl_2$ and HCl .The specimens were then rinsed in deionized water and activated through immersion in an activator containing $PdCl_2$ and HCl . The substrates were then rinsed in a large volume of deionized water for 10 min to prevent contamination the plating bath. The electroless plating process carried out immediately after activation. Then all samples immersed in the electroless bath containing copper sulfate, nickel sulfate, sodium hypophosphite, sodium citrate, boric acid and potassium ferrocyanide. In the post-treatment stage, the Cu-Ni-P plated cotton fabric samples were rinsed with deionized water, ethylalcohol at home temperature for 20 min immediately after the metallising reaction of electroless Cu-Ni-P plating.Then the plated sample dried in oven at 70°C.

The weights (g) of fabric specimens with the size of 100 mm × 100 mm square before and after treatment were measured by a weight meter (HR200, AND Ltd., Japan). The percentage for the weight change of the fabric is calculated in equation (1).

$$I_W = \frac{W_f - W_0}{W_0} \times 100\% \qquad (1)$$

where I_w is the percentage of increased weight, W_f is the final weight after treatment, W_o is the original weight.

The thickness of fabric before and after treatment was measured by a fabric thickness tester (M034A, SDL Ltd., England) with a pressure of 10 g/cm^2. The percentage of thickness increment were calculated in accordance to equations (2) .

$$T_I = \frac{T_F - T_0}{T_0} \times 100\% \qquad (2)$$

where T_I is the percentage of thickness increment, T_f is the final thickness after treatment, T_o is the original thickness.

A Bending Meter (003B, SDL Ltd., England) was employed to measure the degree of bending of the fabric in both warp and weft directions.The flexural rigidity of fabric samples expressed in Ncm is calculated in equation (3).

$$G = W \times C^3 \qquad (3)$$

where G (N-cm) is the average flexural rigidity, W (N/cm^2) is the fabric mass per unit area, C(cm) is the fabric bending length.

The dimensional changes of the fabrics were conducted to assess shrinkage in length for both warp and weft directions and tested with(M003A, SDL Ltd., England) accordance with stansdard testing method (BS EN 22313:1992). The degree of shrinkage in length expressed in percentage for both warp and weft directions were calculated according in accordance to equation (4).

$$D_c = \frac{D_f - D_0}{D_0} \times 100 \qquad (4)$$

where D_c is the average dimensional change of the treated swatch, D_o is the original dimension, D_f is the final dimension after laundering.

Tensile properties and elongation at break were measured with standard testing method ISO 13934-1:1999 using a Micro 250 tensile tester.

Color changing under different application conditions for two standard testing methods, namely, (1) ISO 105-C06:1994 (color fastness to domestic and commercial laundering, (2) ISO 105-A02:1993 (color fastness to rubbing) were used for estimate.

Scanning electron microscope (SEM, XL30 PHILIPS) was used to characterize the surface morphology of deposits. WDX analysis(3PC, Microspec Ltd., USA) was used to exist metallic particles over surface Cu-Ni-P alloy plated cotton fabrics. The chemical composition of the deposits was determined using X-ray energy dispersive spectrum(EDS) analysis attached to the SEM.

The coaxial transmission line method as described in ASTM D 4935-99 was used to test the EMI shielding effectiveness of the conductive fabrics.The set-up consisted of a SE tester, which was connected to a spectrum analyzer. The frequency is scanned from 50 MHz to 2.7 GHz are taken in transmission. The attenuation under transmission were measured equivalent to the SE.

RESULTS AND DISCUSSION

1. Energy Dispersive Elemental Analysis

The composition of the deposits was investigated using X-ray energy dispersive spectrum(EDS) elemental analysis. The deposits consisted mainly of copper with small amounts of nickel and phosphorus. Table 1 shows the weight pecent of all detectd elemental analysis.

Table 1. Elemental analysis of electroless copper plated using hypophosphite and nickel ions

Element	Copper	Nickel	Phosphorous
~ wt%	96.5	3	0.5

The nickel and phosphorus atom s in the copper lattice possibly increase the crystal defects in the deposit. Moreover, as non-conductor, phosphorus will make the electrical resistivity of the deposits higher than pure copper. Electrless plating of copper conductive layer on fabric surface employs hypophosphite ion to reduce copper ion to neural copper particle. However the reduction process extremely accelersts by addition of Ni^{2+}. Addition of

Ni^{2+} also sediments tiny amount of nickel and phosphourus elements. Follwing formulations show the mechanism of copper electroless plating using hypophosphite.

$$2H_2PO_2^- + Ni^{2+} + 2OH^- \rightleftharpoons Ni^0 + 2H_2PO_3^- + H_2$$

$$2H_2PO_2^- + 2\ OH^- \xrightarrow[\text{surface}]{\text{Ni}} 2e^- + 2H_2PO_3^- + H_2$$

$$Cu^{2+} + 2e^- \rightleftharpoons Cu^0$$

$$Ni^0 + Cu^{2+} \rightleftharpoons Ni^{2+} + Cu^0$$

2. Fabric Weight and Thickness

The change in weight and thickness of the untreated cotton and Cu-Ni-P alloy plated cotton fabrics are shown in Table 1.

Table 2. Weight and thickness of the untreated and Cu-Ni-P-plated cotton fabrics

Specimen (10 cm× 10 cm)	Weight (g)	Thickness(mm)
Untreated cotton	2.76	0.4378
Cu-Ni-P plated cotton	3.72 (↑18.47 %)	0.696(↑5.7 %)

The results presented that the weight of chemically induced Cu-Ni-P-plated cotton fabric was heavier than the untreated one. The measured increased percentages of weight was 18.47% .This confirmed that Cu-Ni-P alloy had clung on the surface of cotton fabric effectively. In the case of thickness measurement, the cotton fabric exhibited a 5.7 % increase after being subjected to metallization.

3. Fabric Bending Rigidity

Fabric bending rigidity is a fabric flexural behavior that is important for evaluating the handling of the fabric. The bending rigidity of the untreated cotton and Cu-Ni-P-plated cotton fabrics is shown in Table 2.

Table 3. Bending rigidity of the untreated and Cu-Ni-P-plated cotton fabrics

Specimen	Bending (N·cm)	
	warp	weft
Untreated cotton	1	0.51
Cu-Ni-P plated cotton	1.17(↑11.39%)	0.66(↑30.95%)

The results proved that the chemical plating solutions had reacted with the original fabrics during the entire process of both acid sensitization and alkaline plating treatment .

After electroless Cu-Ni-P alloy plating, the increase in bending rigidity level of the Cu-Ni-P-plated cotton fabrics was estimated at 11.39 % in warp direction and 30.95 % in weft direction respectively. The result of bending indicated that the Cu-Ni-P-plated cotton fabrics became stiffer handle than the untreated cotton fabric.

4. Fabric Shrinkage

The results for the fabric Shrinkage of the untreated cotton and Cu-Ni-P-plated cotton fabrics are shown in Table 3.

Table 4. Dimensional change of the untreated and Cu-Ni-P-plated cotton fabrics

Specimen	Shrinkage (%)	
	warp	weft
Untreated cotton	0	0
Cu-Ni-P plated cotton	-8	-13.3

The measured results demonstrated that the shrinkage level of the Cu-Ni-P-plated cotton fabric was reduced by 8% in warp direction and 13.3% in weft direction respectively.

After the Cu-Ni-P-plated, the copper particles could occupy the space between the fibres and hence more copper particles were adhered on the surface of fibre. Therefore, the surface friction in the yarns and fibres caused by the Cu-Ni-P particles could then be increased. When compared with the untreated cotton fabric, the Cu-Ni-P-plated cotton fabrics shown a stable structure.

5. Fabric Tensile Strength and Elongation

The tensile strength and elongation of cotton fabrics was enhanced by the electroless Cu-Ni-P alloy plating process as shown in Table 4.

Table 5. Tensile strength and percentage of elongation at break load of the untreated and Cu-Ni-P- Fabrics plated cotton fabrics

Specimen	Percentage of elongation (%)		Breaking load (N)	
	warp	weft	warp	weft
Untreated cotton	6.12	6.05	188.1	174.97
Cu-Ni-P plated cotton	6.98(↑12.5%)	6.52(↑7.8%)	241.5(↑28.4%)	237.3(↑35.62%)

The metallized cotton fabrics had a higher breaking load with a 28.44% increase in warp direction and a 35.62% increase in weft direction than the untreated cotton fabric. This was due to the fact that more force was required to pull the additionalmetal-layer coating.

The results of elongation at break were 12.5 % increase in warp direction and 7.8 % increase in weft direction, indicating that the Cu-Ni-P-plated fabric encountered little change when compared with the untreated cotton fabric. This confirmed that with the metallizing treatment, the specimens plated with metal particles was demonstrated a higher frictional force of fibers. In addition, the deposited metal particles which developed a linkage force to hamper the movement caused by the applied load.

6. Color Change Assessment

The results of evaluation of color change under different application conditions, washing, rubbing are shown in Table 5.

Table 6. Washing and rubbing fastness of the untreated and Cu-Ni-P Fabrics plated cotton fabrics

Specimen	Washing	Rubbing	
		Dry	Wet
Cu-Ni-P plated cotton	5	4-5	3-4

The results of the washing for the Cu-Ni-P-plated cotton fabric was grade 5 in color change. This confirms that the copper particles had good performance during washing. The result of the rubbing fastness is shown in Table 5. According to the test result, under dry rubbing condition, the degree of staining was recorded to be grade 4-5, and the wet rubbing fastness showed grade 3-4 in color chang . This results showed that the dry rubbing fastness had a lower color change in comparison with the wet crocking fastness. In view of the overall results, the rubbing fastness of the Cu-Ni-P-plated cotton fabric was relatively good when compared with the commercial standard

7. Surface Morphology

Scanning electron microscopy (SEM) of the untreated and Cu-Ni-P-plated cotton fabric is shown in Figures 1 with magnification of 250x. Microscopic evidents of copper coated fabrics shows the formation of evenness copper particles on fabric surface and structure.

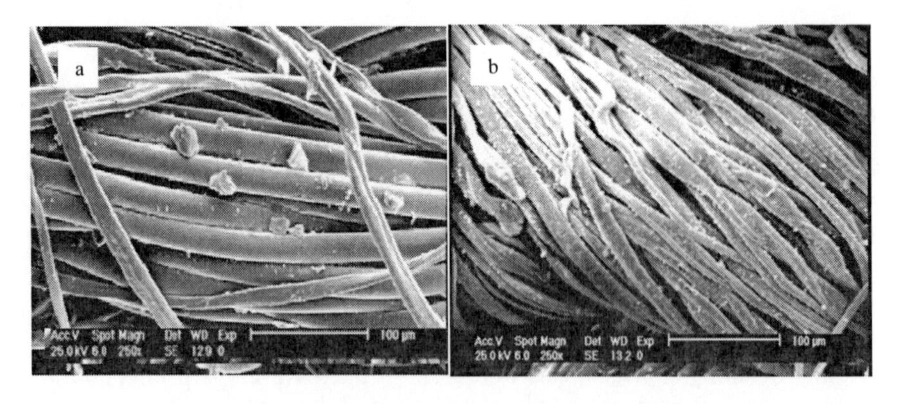

Figure 1.SEM photographs of the (a)untreated cotton fabric (b) Cu-Ni-P plated cotton fabric.

Figure 2 showes the SEM and WDX analysis copper plated surfaces of cotton fiber.It was observed that the cotton fibers surface was covered by Cu-Ni-P alloy particles composing of an evenly distributed mass. In addition,WDX analysis indicated that the deposits became more compact , uniform and smoother also exist homogenous metal particle distribution over coated fabricsurface. These results indicate that the effect of chemical copper plating is sufficient and effective to provide highly conductive surface applicable for EMI shielding use.

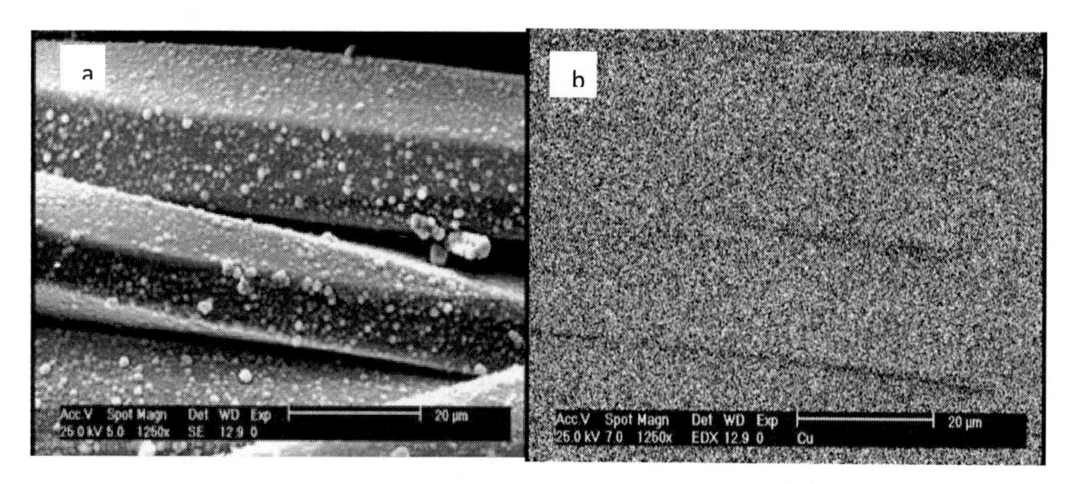

Figure2.(a)SEM photograph of the Cu-Ni-P plated cotton fabric (b) WDX analysis of the Cu-Ni-P plated cotton fabric.

8. Shielding Effectiveness

Electromagnetic shielding means that the energy of electromagnetic radiation is attenuated by reflection or absorption of an electromagnetic shielding material, which is one of the effective methods to realize electromagnetic compatibility .The unit of EMISE is given in decibels (dB). The EMI shielding effectiveness value was calculated from the ratio of the incident to transmitted power of the electromagnetic wave in the following equation :

$$SE = 10\log\left|\frac{P_1}{P_2}\right| = 20\left|\frac{E_1}{E_2}\right| \qquad (5)$$

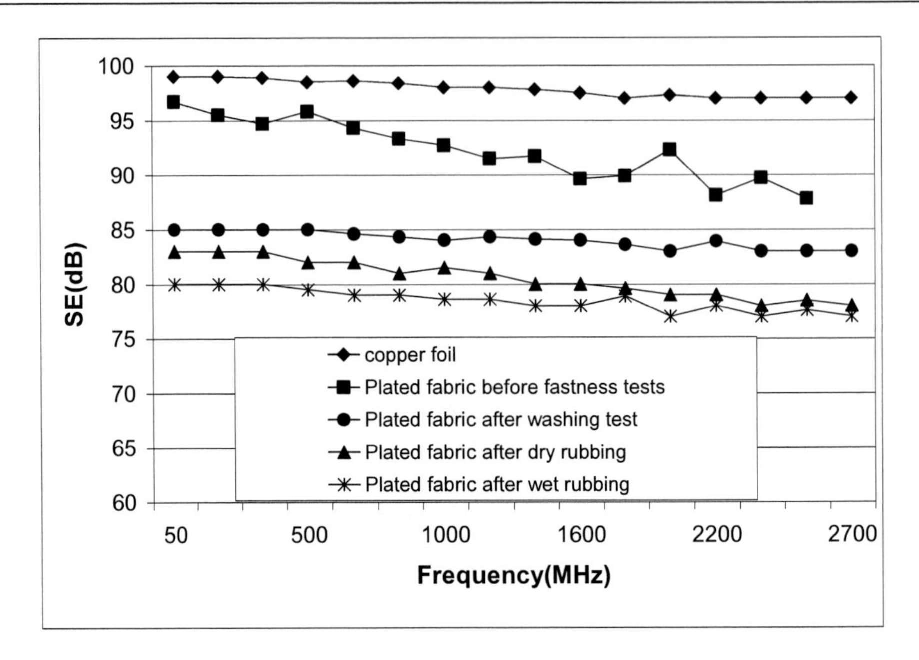

Figure3. The shielding effectiveness of various conductive sample

where P_1 (E_1) and P_2 (E_2) are the incident power (incident electric field) and the transmitted power (transmitted electric field), respectively. Figure 3 indicates the shielding effectiveness (SE) of the copper-coated fabrics with 1 ppm $K_4Fe(CN)_6$ comapred to copper foil and other sample after washing and rubbing fastness test. The shielding effectivness test applied on five different conductive sample including copper foil, electroless plateed of Cu-Ni-P alloy particle on cotton fabric, Electroless plated fabric after washing test, Electroless plated fabric after dry and wet rubbing. As it can be expected copper foil with completely metallic structure shows the best shielding effectivness performance according to higher conductivity compared to other conductive fabric sample. However, SE of copper-coated cotton fabric was above 90 dB and the tendency of SE kept similarity at the frequencies 50 MHz to 2.7 GHz . The aquired results for samples after washing fastness test shows nearly 10 % decrease over frequency range which is still applicable for practical EMI shielding use. Two other samples after rubbing shows respectively 12% and 15% reduction in shielding effectiveness value but the presented results still show an accepted level of shielding around 80 dB. The SE reduction after fastness tests is a quite normal behavior which is likely due to removing of conductive particles from fabric surface. However, the compact and homogenous distribution of conductive particles provide a great conductive coating on fabric surface with high durability even after wahing or rubbing tests. The copper-coated cotton fabric has a practical usage for many EMI shielding application requirements.

CONCLUSION

In this study, Electroless plating of Cu-Ni-P alloy process onto cotton fabrics was demonstrated. Both uncoated and Cu-Ni-P alloy coated cotton fabrics were evaluated with measurement weight change, fabric thickness, bending rigidity, fabric shrinkage, tensile strength , percentage of elongation at break load and color change assessment. The results showed significant increase in weight and thickness of chemically plated cotton fabric. Coated samples showed better properties and stable structure with uniformly distributed metal particles. The SE of copper-coated cotton fabric was above 90 dB and the tendency of SE kept similarity at the frequencies 50 MHz to 2.7GHz.Also, the evaluaton of SE after standard washing and abration confirms the supreme durable shielding behavior. The copper-coated cotton fabric has a practical usage for many EMI shielding application requirements.

REFERENCES

[1] R-H Guo , S-Q Jiang , C-W-M Yuen ,M-C-F Ng , An alternative process for electroless copper plating on polyester fabric, *J Mater Sci: Mater Electron,* DOI 10.1007/s10854-008-9594-4 (2008).

[2] Y-M Lin, S-H Yen, Effects of additives and chelating agents on electroless copper plating , *Appl. Surf. Sci.,*178(1-4),116-126(2001)

[3] G- Xueping ,W- Yating ,L- Lei , Sh- Bin, H- Wenbin , Electroless plating of Cu–Ni–P alloy on PET fabrics and effect of plating parameters on the properties of conductive fabrics, *Journal of Alloys and Compounds,* 455, 308–313,(2008).

[4] J-Li, H-Hayden, P-A- Kohl, The influence of 2,2'-dipyridyl on non-formaldehyde electroless copper plating, *Electrochim Acta* 49 , 1789–1795, (2004).

[5] H- Larhzil, M- Cisse, R- Touir, M- Ebn Touhami, M- Cherkaoui, Electrochemical and SEM investigations of the influence of gluconate on the electroless deposition of Ni–Cu–P alloys , *Electrochimica Acta* 53 ,622–628, (2007).

[6] J-G Gaudiello,G-L Ballard, Mechanistic insights into metal-mediated electroless copper plating Employing hypophosphite as a reducing agent ,*IBM J. RES. DEVELOP*, 37 (2) ,(1993).

[7] G-Xueping ,W-Yating ,L- Lei , Sh-Bin, H-Wenbin, Electroless copper plating on PET fabrics using hypophosphite as reducing agent, *Surface & Coatings Technology* 201, 7018–702, (2007).

[8] E-G Han, E-A Kim, K-W Oh, Electromagnetic interference shilding effectiveness of electroless Cu-platted PET fabrics ,*Synth. Met.* 123 , 469–476, (2001).

[9] S- Q Jiang, R- H Guo, Effect of Polyester Fabric through Electroless Ni-P Plating, *Fibers and Polymers*, 9 (6), 755-760,(2008).

[10] S- S Djokic, *Fundamental Aspects of Electrometallurgy: Chapter 10, Metal Deposition without an External Current*, Eds., K. I. Popov, S. S. Djokic B. N. Grgur, pp. 249-270, Kluwer Academic Publishers, New York, 2002.

[11] O- Mallory, B- J Hajdu, Eds., *"Eleolesctrs Plating: Fundamentals and Applications"*, Noyes Publication, New York, 1990.

[12] R-C. Agarwala, V- Agarwala, *"Electroless alloy/composite coatings: a review"*, *Sadhana*, 28, 475- 493, (2003).

[13] M. Paunovic, *"Electrochemical Aspects of Electroless Deposition of Metals", Plating*, 51, 1161-1167, 1968.

[14] H-F Chang, W-H Lin, TPR study of electroless plated copper catalysts, *Korean J. Chem. Eng.*, 15(5), 559-562 (1998) LYSTS

Chapter 8

SYSTEMATIC PARAMETERS IN FORMATION OF ELECTROSPUN NANOFIBERS

1. INTRODUCTION AND BACKGROUND

Nanotechnology has become in recent years a topic of great interest to scientists and engineers, and is now established as prioritized research area in many countries. The reduction of the size to the nano-meter range brings an array of new possibilities in terms of material properties, in particular with respect to achievable surface to volume ratios. Electrospinning of nanofibers is a novel process for producing superfine fibers by forcing a solution through a spinnerette with an electric field. An emerging technology of manufacturing of thin natural fibers is based on the principle of electrospinning process. In conventional fiber spinning, the mechanical force is applied to the end of a jet. Whereas in the electrospinnig process the electric body force act on element of charged fluid. Electrospinning has emerged as a specialized processing technique for the formation of sub-micron fibers (typically between 100 nm and 1 μm in diameter), with high specific surface areas. Due to their high specific surface area, high porosity, and small pore size, the unique fibers have been suggested for wide range of applications. Electrospinning of nano fibers offers unique capabilities for producing novel natural nanofibers and fabrics with controllable pore structure.

Some 4-9% of cotton fiber is lost at textile mill in so-called opening and cleaning, which involves mechanically separating compressed clumps of fibers for removal of trapped debris. Another 1 % is lost in drawing and roving-pulling lengths of fiber into longer and longer segments, which are then twisted together for strength. An average of 20% is lost during combing and yarn production. Typically, waste cotton is used in relatively low-value products such as cotton balls, yarn, and cotton batting. A new process for electrospinning waste cotton using a less harmful solvent has been developed.

Electrospinning is an economical and simple method used in the preparation of polymer fibers. The fibers prepared via this method typically have diameters much smaller than is possible to attain using standard mechanical fiber-spinning technologies.[1] Electrospinning has gain much attention in the last few years as a cheap and straightforward method to produce nanofibers. Electrospinning differs from the traditional wet/dry fiber spinning in a number of ways, of which the most striking differences are the origin of the pulling force and the final fiber diameters. The mechanical pulling forces in the traditional industrial fiber spinning processes lead to fibers in the micrometer range and are contrasted in electrospinning by electrical pulling forces that enable the production of nanofibers. Depending on the solution properties, the throughput of single-jet electrospinning systems ranges around 10 ml/min. This low fluid throughput may limit the industrial use of electrospinning. A stable cone-jet mode followed by the onset of the characteristic bending instability, which eventually leads to great reduction in the jet diameter, necessitate the low flow rate [2]. When the diameters of cellulose fiber materials are shrunk from micrometers (e.g. 10–100 mm) to submicrons or nanometers, there appear several amazing characteristics such as very large surface area to volume ratio (this ratio for a nanofiber can be as large as 103 times of that of a microfiber), flexibility in surface functionalities, and superior mechanical performance (e.g. stiffness and tensile strength) compared with any other known form of the material.

These outstanding properties make the polymer nanofibers to be optimal candidates for many important applications [3]. These include filter media, composite materials, biomedical applications (tissue engineering scaffolds, bandages, drug release systems), protective clothing for the military, optoelectronic devices and semi-conductive materials, biosensor/chemosensor [4]. Another biomedical application of electrospun fibers that is currently receiving much attention is drug delivery devices. Researchers have monitored the release profile of several different drugs from a variety of biodegradable electrospun membranes. Other application for electrospun fibers is porous membranes for filtration devices. Due to the inter-connected network type structure that electrospun fibers form; they exhibit good tensile properties, low air permeability, and good aerosol protection capabilities. Moreover, by controlling the fiber diameter, electrospun fibers can be produced over a wide range of porosities. Research has also focused on the influence of charging effects of electrospun non-woven mats on their filtration efficiency. The filtration properties slightly depended on the surface charge of the membrane, however the fiber diameter was found to have the strongest influence on the aerosol penetration. Electrospun fibers are currently being utilized for several other applications as well. Some of these include areas in nanocomposites. Figure 1 compares the dimensions of nanofibers, micro fibers and ordinary fibers. These When the diameters of polymer fiber materials are shrunk from micrometers (for example 10-100 μm) to sub-microns or nano meters (for example 10 $\times 10^{-3}$-100$\times 10^{-3}$ μm), there appear several amazing characteristics such as very large surface area to volume ratio, flexibility in surface functionalities, and superior mechanical performance compared with any other known form of material. These outstanding properties make the polymer nanofibers to be optimal candidates for many important applications [4].

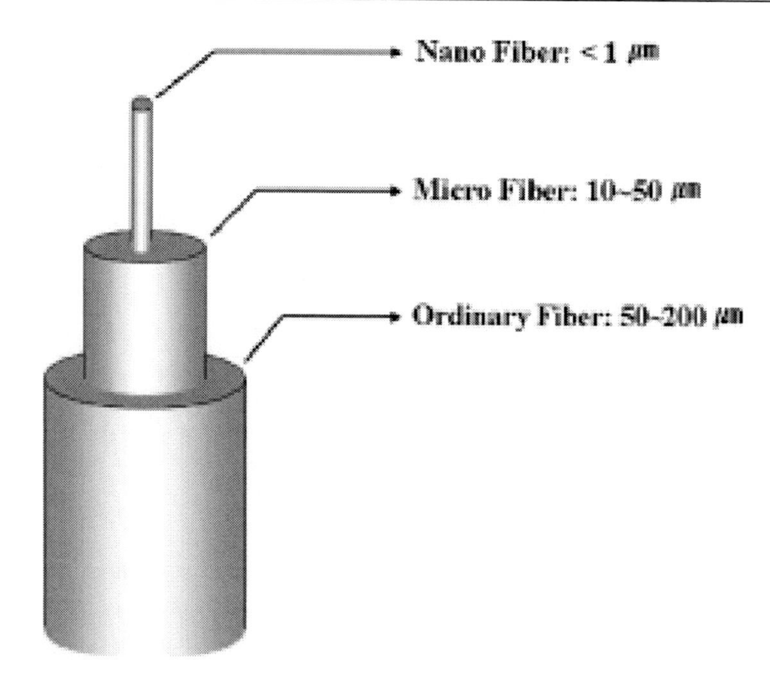

Figure 1.Classifications of fibers by the fiber diameter.

1.1. Electrospinning Set Up

A schematic diagram to interpret electrospinning of nanofibers is shown in Figure 2.

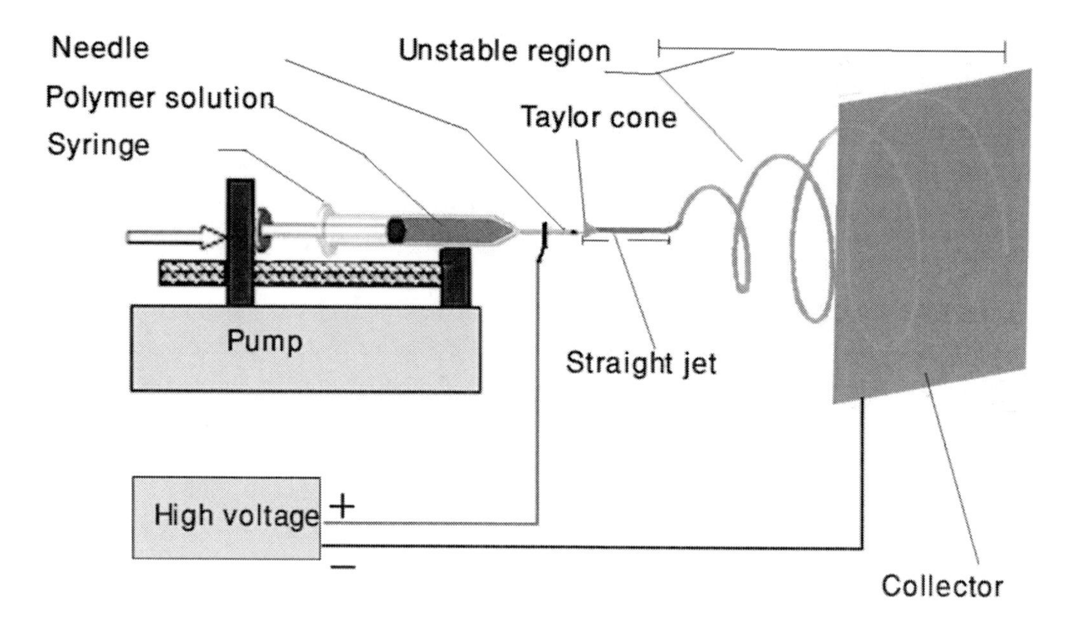

Figure 2. Shematic of electrospinning set up.

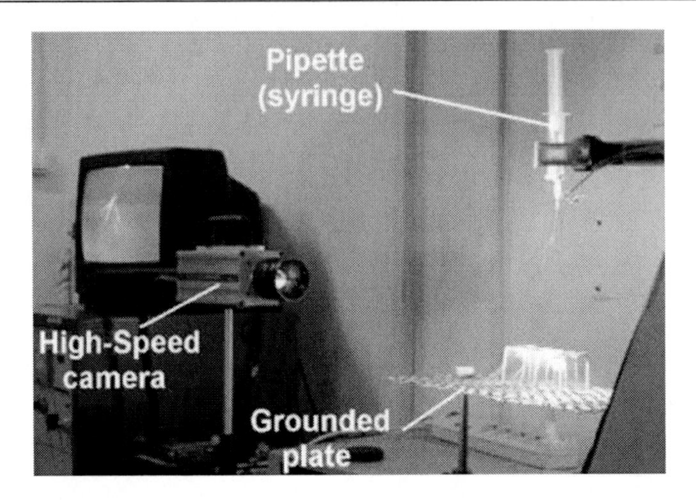

Figure 3. Electrospinning process.

There are basically three components to fulfill the process: a high voltage supplier, a capillary tube with a pipette or needle of small diameter, and a metal collecting screen. In the electrospinning process (Figure 3) a high voltage is used to create an electrically charged jet of polymer solution or melt out of the pipette. Before reaching the collecting screen, the solution jet evaporates or solidifies, and is collected as an interconnected web of small fibers [5]. One electrode is placed into the spinning solution/melt or needle and the other attached to the collector. In most cases, the collector is simply grounded. The electric field is subjected to the end of the capillary tube that contains the solution fluid held by its surface tension. This induces a charge on the surface of the liquid. Mutual charge repulsion and the contraction of the surface charges to the counter electrode cause a force directly opposite to the surface tension [6]. As the intensity of the electric field is increased, the hemispherical surface of the fluid at the tip of the capillary tube elongates to form a conical shape known as the Taylor cone [7]. Further increasing the electric field, a critical value is attained with which the repulsive electrostatic force overcomes the surface tension and the charged jet of the fluid is ejected from the tip of the Taylor cone [8]. The jet exhibits bending instabilities due to repulsive forces between the charges carried with the jet. The jet extends through spiraling loops, as the loops increase in diameter the jet grows longer and thinner until it solidifies or collects on the target [9]. A schematic diagram to interpret electrospinning of polymer nanofibers is shown in Figure 4 and Figure 5. There are basically three components to fulfill the process: a high voltage supplier, a capillary tube with a pipette or needle of small diameter, and a metal collecting screen. In the electrospinning process a high voltage is used to create an electrically charged jet of polymer solution or melt out of the pipette. Before reaching the collecting screen or drum, the solution jet evaporates or solidifies, and is collected as an interconnected web of small fibers. When an electric field is applied between a needle capillary end and a collector, surface charge is induced on a polymer fluid deforming a spherical pendant droplet to a conical shape. As the electric field surpasses a threshold value where electrostatic repulsion force of surface charges overcome surface tension, the charged fluid jet is ejected from the tip of the Taylor cone and the charge density on the jet interacts with the external field to produce instability.

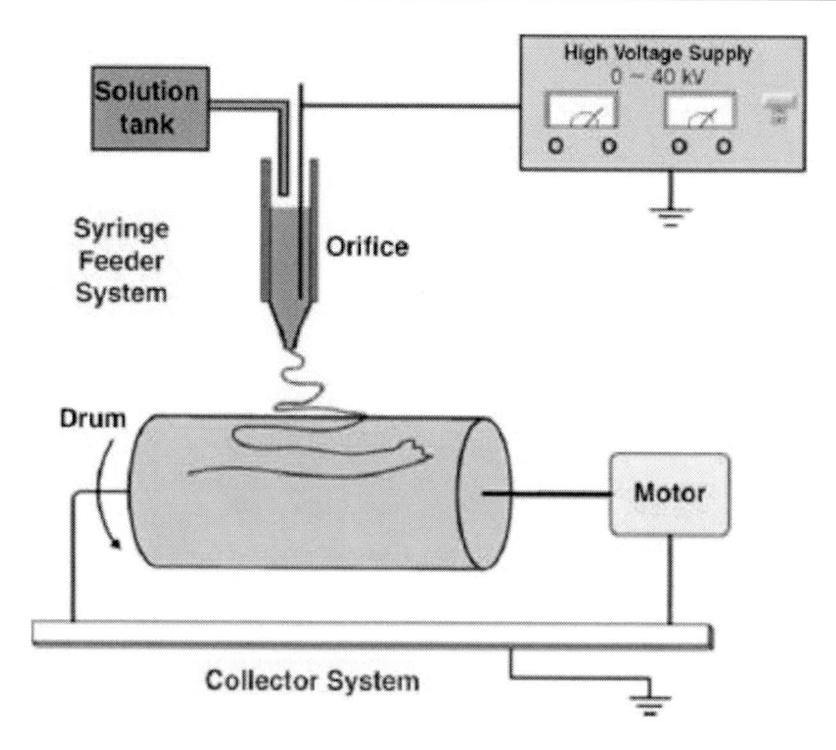

Figure 4.Schematic diagram of the electrospinning set up by drum collector.

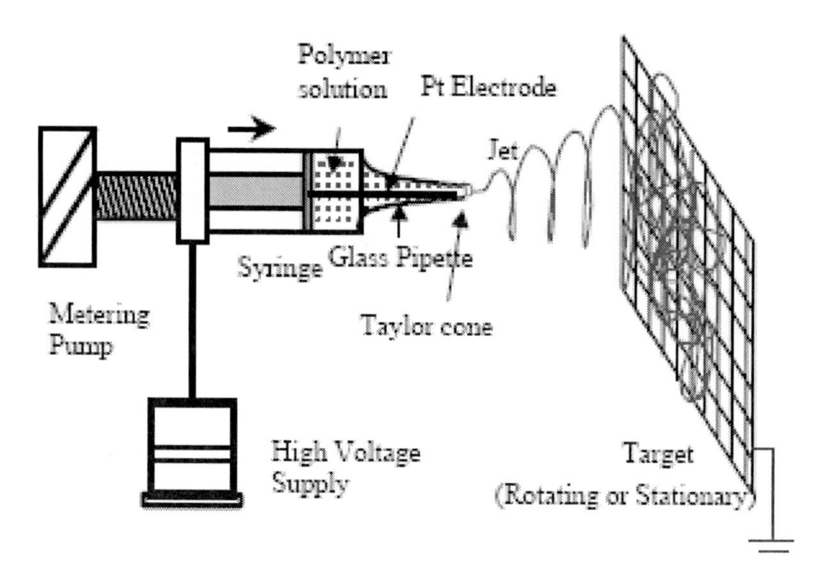

Figure 5.Schematic diagram of the electrospinning set up by screen collector.

It has been found that morphology such as fiber diameter and its uniformity of the electrospun polymer fibers are dependent on many processing parameters. These parameters can be divided into three groups as shown in Table 1. Under certain condition, not only uniform fibers but also beads-like formed fibers can be produced by electrospinning. Although the parameters of the electrospinning process have been well analyzed in each of polymers these information has been inadequate enough to support the electrospinning of ultra-fine nanometer scale polymer fibers. A more systematic parametric study is hence required to investigate.

Table 1. Processing parameters in electrospinning

Solution properties	Viscosity
	Polymer concentration
	Molecular weight of polymer
	Electrical conductivity
	Elasticity
	Surface tension
Processing conditions	Applied voltage
	Distance from needle to collector
	Volume feed rate
	Needle diameter
Ambient conditions	Temperature
	Humidity
	Atmospheric pressure

2. Effect of Systematic Parameters on Electrospun Nanofibers

It has been found that morphology such as fiber diameter and its uniformity of the electrospun nanofibers are dependent on many processing parameters. These parameters can be divided into three main groups: a) solution properties, b) processing conditions, c) ambient conditions. Each of the parameters has been found to affect the morphology of the electrospun fibers.

2.1. Solution Properties

Parameters such as viscosity of solution, solution concentration, molecular weight of solution, electrical conductivity, elasticity and surface tension, have important effect on morphology of nanofibers.

2.1.1. Viscosity

The viscosity range of a different nanofiber solution which is spinnable is different. One of the most significant parameters influencing the fiber diameter is the solution viscosity. A higher viscosity results in a large fiber diameter. Figure 6 shows the representative images of beads formation in electrospun nanofibers. Beads and beaded fibers are less likely to be formed for the more viscous solutions. The diameter of the beads become bigger and the average distance between beads on the fibers longer as the viscosity increases.

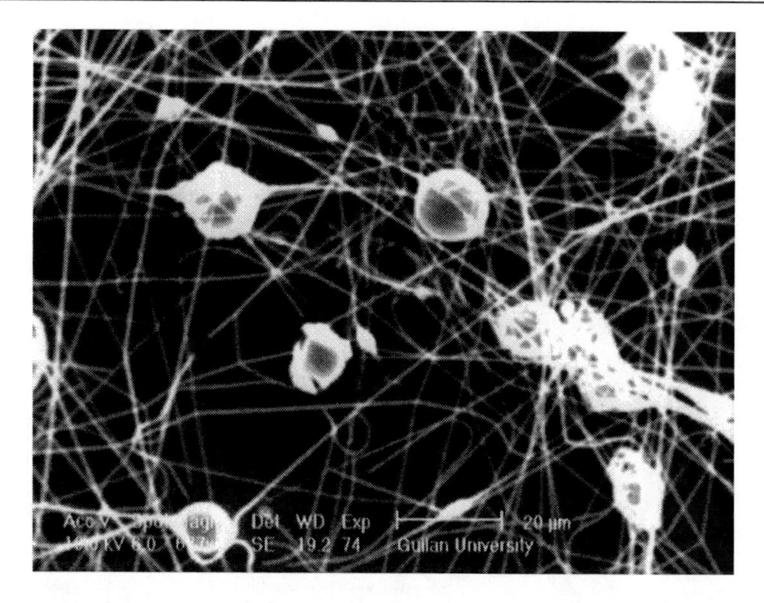

Figure 6. Electron micrograph of beads formation in electrospun nanofibers.

2.1.2. Solution Concentration

In electrospinning process, for fiber formation to occur, a minimum solution concentration is required. As the solution concentration increase, a mixture of beads and fibers is obtained (Figure 7). The shape of the beads changes from spherical to spindle-like when the solution concentration varies from low to high levels. It should be noted that the fiber diameter increases with increasing solution concentration because the higher viscosity resistance. Nevertheless, at higher concentration, viscoelastic force which usually resists rapid changes in fiber shape may result in uniform fiber formation. However, it is impossible to electrospin if the solution concentration or the corresponding viscosity become too high due to the difficulty in liquid jet formation.

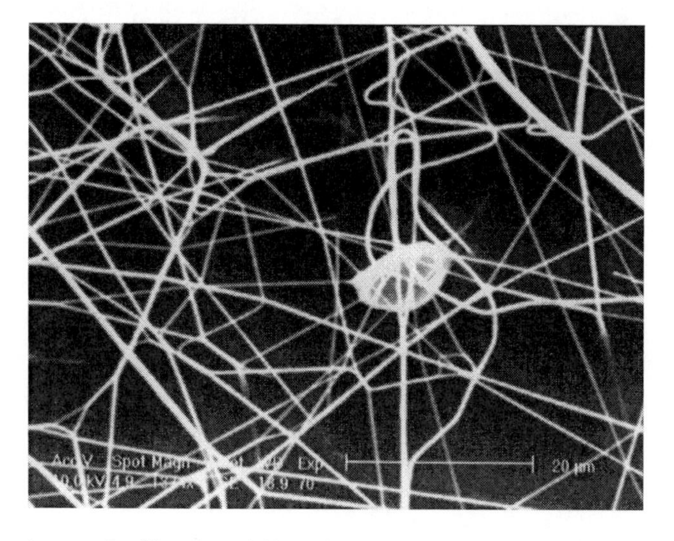

Figure 7. Electron micrograph of beads and fibers formation in electrospun nanofibers.

2.1.3. Molecular Weight

Molecular weight also has a significant effect on the rheological and electrical properties such as viscosity, surface tension, conductivity and dielectric strength. It has been reported that too low molecular weight solution tend to form beads rather than fibers and high molecular weight nanofiber solution give fibers with larger average diameter (Figure 8).

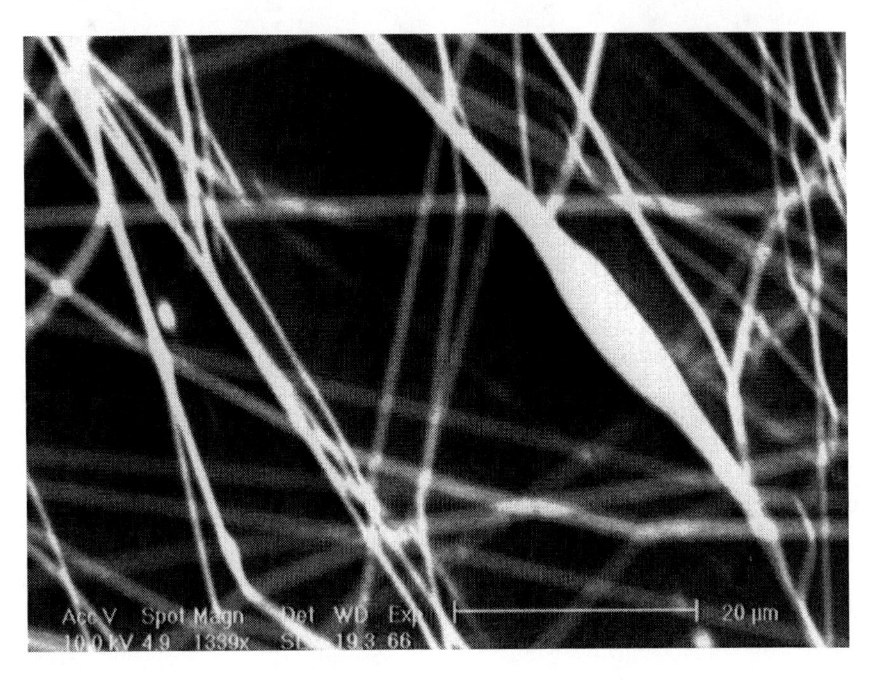

Figure 8. Electron micrograph of variable diameter formation in electrospun nanofibers

2.1.4. Surface Tension

The surface tension of a liquid is often defined as the force acting at right angles to any line of unit length on the liquid surface. However, this definition is somewhat misleading, since there is no elastic skin or tangential force as such at the surface of a pure liquid, It is more satisfactory to define surface tension and surface free energy as the work required to increase the area of a surface isothermally and reversibly by unit amount. As a consequence of surface tension, there is a balancing pressure difference across any curved surface, the pressure being greater on the concave side. By reducing surface tension of a nanofiber solution, fibers could be obtained without beads (Figures 9-10). This might be correct in some sense, but should be applied with caution. The surface tension seems more likely to be a function of solvent compositions, but is negligibly dependent on the solution concentration. Different solvents may contribute different surface tensions. However, not necessarily a lower surface tension of a solvent will always be more suitable for electrospinning. Generally, surface tension determines the upper and lower boundaries of electrospinning window if all other variables are held constant. The formation of droplets, bead and fibers can be driven by the surface tension of solution and lower surface tension of the spinning solution helps electrospinning to occur at lower electric field.

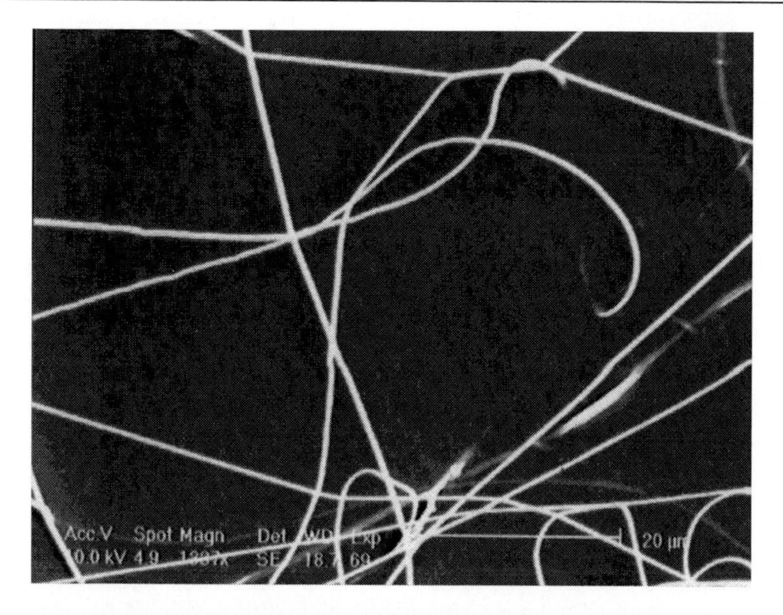

Figure 9. Electron micrograph of electrospun nanofiber without beads formation.

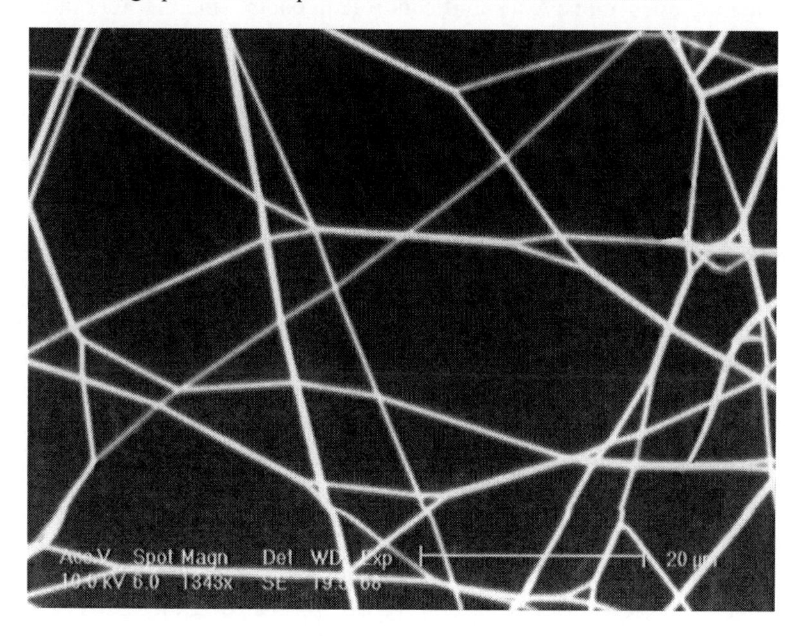

Figure 10. Electron micrograph of electrospun nanofiber without beads formation.

2.1.5. Solution Conductivity

There is a significant drop in the diameter of the electrospun nanofibers when the electrical conductivity of the solution increases. Beads may also be observed due to low conductivity of the solution, which results in insufficient elongation of a jet by electrical force to produce uniform fiber. In general, electrospun nanofibers with the smallest fiber diameter can be obtained with the highest electrical conductivity. This interprets that the drop in the size of the fibers is due to the increased electrical conductivity.

2.2. Processing Condition

2.2.1. Applied Voltage

In the case of electrospinning, the electric current due to the ionic conduction of charge in the nanofiber solution is usually assumed small enough to be negligible. The only mechanism of charge transport is the flow of solution from the tip to the target. Thus, an increase in the electrospinning current generally reflects an increase in the mass flow rate from the capillary tip to the grounded target when all other variables (conductivity, dielectric constant, and flow rate of solution to the capillary tip) are held constant.

With the increase of the electrical potential the resulting nanofibers became rougher. It is sometimes reported that a diameter of electrospun fibers does not significantly affected by an applied voltage This voltage effects is particularly diminished when the solution concentration is low. Applied voltage may affect some factors such as mass of solution fed out from a tip of needle, elongation level of a jet by an electrical force, morphology of a jet (a single or multiple jets), etc. A balance among these factors may determine a final diameter of electrospun fibers. It should be also noted that beaded fibers may be found to be electrospun with too high level of applied voltage. Although voltage effects show different tendencies, but the voltage generally does not have a significant role in controlling the fiber morphology.

Nevertheless, increasing the applied voltage (*i.e.,* increasing the electric field strength) will increase the electrostatic repulsive force on the fluid jet which favors the thinner fiber formation. On the other hand, the solution will be removed from the capillary tip more quickly as jet is ejected from Taylor cone. This results in the increase of the fiber diameter.

2.2.2. Feed Rate

The morphological structure can be slightly changed by changing the solution flow rate as shown in Figure 11. At the flow rate of 0.3 ml/h, a few of big beads were observed on the fibers. When the flow rate exceeded a critical value, the delivery rate of the solution jet to the capillary tip exceeds the rate at which the solution was removed from the tip by the electric forces. This shift in the mass-balance resulted in sustained but unstable jet and fibers with big beads formation.

The solution's electrical conductivity, were found as dominant parameters to control the morphology of electrospun nanofibers [9]. In the case of low-molecular- weight liquid, when a high electrical force is applied, formation of droplets can occur. A theory proposed by Rayleigh explained this phenomenon. As evaporation of a droplet takes place, the droplet decreases in size. Therefore the charge density of its surface is increased. This increase in charge density due to Coulomb repulsion overcomes the surface tension of the droplet and causes the droplet to split into smaller droplets. However, in the case of a solution with high molecular weight liquid, the emerging jet does not break up into droplets, but is stabilized and forms a string of beads connected by a fiber. As the concentration is increased, a string of connected beads is seen, and with further increase there is reduced bead formation until only smooth fibers are formed. And sometimes spindle-like beads can form due to the extension causing by the electrostatic stress. The changing of fiber morphology can probably be attributed to a competition between surface tension and viscosity. As concentration is increased, the viscosity of the solution increases as well. The surface tension attempts to reduce surface area per unit mass, thereby caused the formation of beads/spheres. Viscoelastic forces resisted the formation of beads and allowed for the formation of smooth fibers.

Therefore formation of beads at lower solution concentration (low viscosity) occurs where surface tension had a greater influence than the viscoelastic force. However, bead formation can be reduced and finally eliminated at higher solution concentration, where viscoelastic forces had a greater influence in comparison with surface tension. But when the concentration is too high, high viscosity and rapid evaporation of solvent makes the extension of jet more difficult, thicker and ununiform fibers will be formed.

Suitable level of processing parameters must be optimized to electrospin solutions into nanofibers with desired morphology and the parameters levels are dependent on properties of solution and solvents used in each of electrospinning process. Understanding of the concept how each of processing parameter affect the morphology of the electrospun nanofibers is essential. All the parameters can be divided into two main groups; i.e. one with parameters which affect the mass of solution fed out from a tip of needle, and the other with parameters which affect an electrical force during electrospinning. Solution concentration, applied voltage and volume feed rate are usually considered to affect the mass. Increased solution concentration and feed rate tend to bring more mass into the jet. High Applied voltage reflects to force to pull a solution out from the needle hence higher applied voltage causes more solution coming out. On the other hand, it should be noted that solution electrical conductivity and applied voltage affect a charge density thus an electrical force, which acts to elongate a jet during electrospinning.

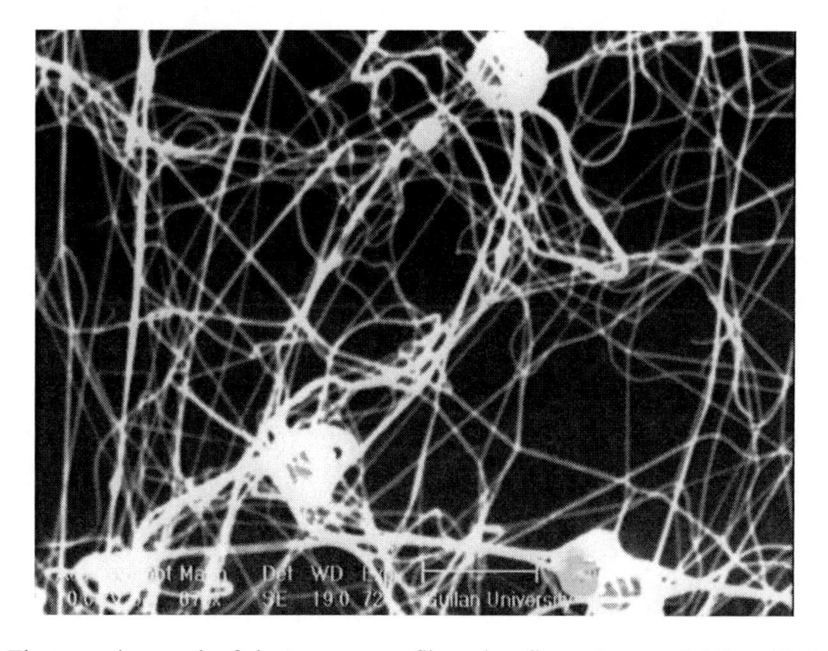

Figure 11. Electron micrograph of electrospun nanofiber when flow rate exceeded the critical value.

3. EXPERIMENTAL

Polyacrylonitrile (PAN) fiber (Dolan) was from Hoechst .NMP (N-methyl-2- pyrolidon) was from Riedel-de Haën. The polyaniline used was synthesized in our laboratory. Polyaniline (PANi) was synthesized by the oxidative polymerization of aniline in acidic media. 3 ml of distilled aniline was dissolved in 150 ml of 1N HCl and kept at 0 °C, 7. 325g of $(NH_4)_2S_2O_8$ was dissolved in 35 ml of 1N HCl and added dropwise under constant stirring to the aniline/HCl solution over a period of 20 mins. The resulting dark green solution was maintained under constant stirring for 4 hrs filtered and washed with methanol and then with water then it was dried before being added to 150 mL of 1N $(NH_4)OH$ solution. After an additional 4 hrs the solution was filtered and a deep blue emeraldine base form of polyaniline was obtained (PANiEB).Then it was dried and crushed into fine powder and then passed trough a 100 mesh.

The polymer solutions were prepared by first dissolving exact amount of PANi in NMP. The PANi was slowly added to the solvent with constant stirring at room temperature. This solution was then allowed to stir for 1 hour in a sealed container. PAN/NMP solution was prepared separately and added dropwise to the well stirred PANi solution and the blend solution was allowed to stir with a mechanical stirrer for an additional 1 hour.

By mixing different solution ratios (0/100, 50/50, 60/40, 75/25) of 5% PANi solution and 20% PAN solution various polymer blend solutions were prepared with the concentration of polyaniline ranging from 5 wt% to 42 wt%. The fiber diameter and polymer morphology of the electrospun polyaniline/polyacrylonitrile NMP solution were determined using an optical microscope Nikon Microphot-FXA. A small section of the non-woven mat was placed on the glass slide and placed on the microscope sample holder. A scanning electron microscope (SEM) Philips XL-30 was used to take the SEM photographs to do more precise characterization. A small section of the web was placed on SEM sample holder and coated with gold (BAL-TEC SCD 005 sputter coater).

4. Result and Discussion

In our first experiment we tried to find out whether electrospinning of PANi pure solution can result in a web formation or not. Without the addition of PAN to PANi dissolved in NMP, no web formation occurred, because the concentration and viscosity of the solution was not high enough to form a stable drop at the end of the needle and just some dispersed drops were formed on the collector. Adding more polyaniline can not increase the solution viscosity and just resulted in gelation of the solution.

Based on these results the blend solution was electrospun the initial results showed that in room temperature fine fibers are formed. In these blends the concentration of PANi ranged from 5 wt% to 42 wt%. The potential difference between the needle tip and the electrode was 20 Kv. Optical microscope photomicrographs (Figure 12) showed that the fibers are formed but they are entangled to each other also the bead forming is observed. In order to get more uniform webs we tried to obtain the webs at the gap between two metal stripes which were placed on the collector plate, and it was seen that (Figure 13) webs got more uniform but yet the entanglement and beads were observed.

For examining the web formation of PAN it was electrospun from NMP at different concentrations. (10% and 15%) webs were formed at voltages between 17 and 20. PAN web showed an excellent webforming behavior and the resulted webs were uniform (Figure 14).

After that optical microscope micrographs confirmed that the fibers are formed; higher concentrations of PANi were used and the resulted webs were examined by SEM for studying their diameter and morphology more precisely. The SEM photomicrograph revealed that the diameter of fibers in non-woven mat ranged from minimum 160 nm to maximum 560 nm, with an average fiber diameter of 358 nm.

It was noticed that by increasing amount of PAN the fiber formation enhanced and more uniform fibers were obtained; also the fiber diameter variation is smaller, it can be related to intrinsic fiber forming behavior of PAN as it is used widely as the base material for producing fibers and yarns. By increasing PANi ratio amount of beading increased and fibers twisted to each other before reaching the collector, in some parts a uniform web was formed. It seemed that the fibers were wet and it is the cause of sticking of fibers together. In 42% of PANi the webforming was not seen and instead we had an entangled bulk with some polymer drops. Actually with increasing the PANi ratio the fiber diameter decreased as it can be noticed clearly from the results indicated in **Table 2**.

Table 2 Fiber diameters in different PANi ratios.

PANi percent (blend ing ratio)	0% (0/100)	20% (50/50)	27% (60/40)
Fiber diameter	445 nm	372 nm	292 nm

By increasing the temperature of electrospinning environment to 75̊c in order to let the solvent evaporate more rapidly the problem of twisted fiber was overcame and more uniform webs and finer nanofibers were formed but yet the beading problem was seen. More research is in progress to enhance the web characterization and decreases the fiber diameter to real nanometer size.

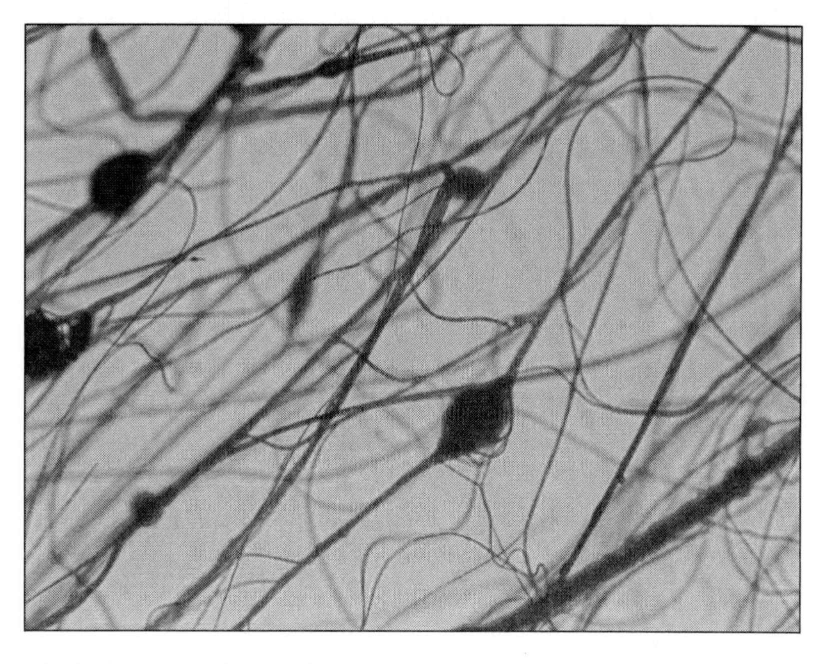

Figure 12. Optical microscope micrograph of 16% PANi blend solution, beads can be observed clearly.

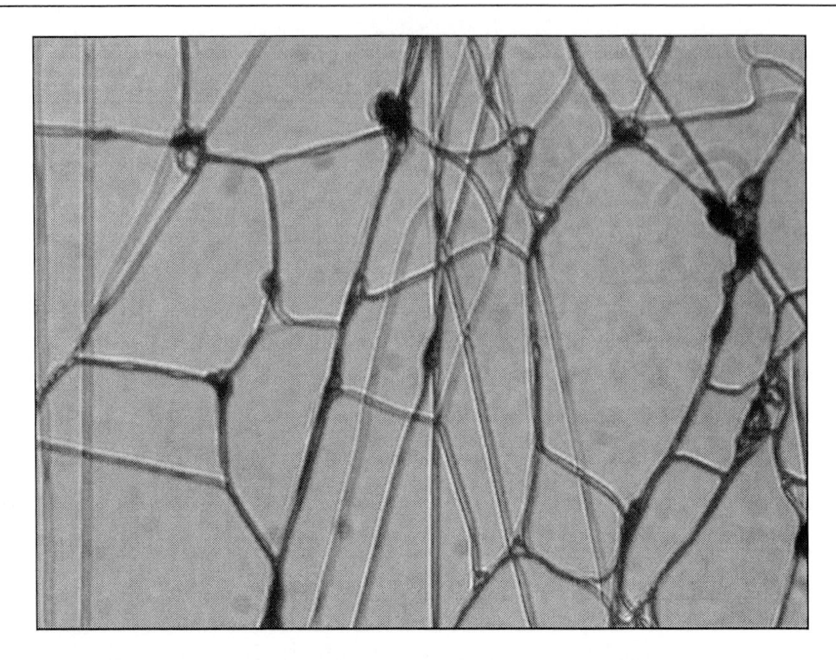

Figure 13. Optical microscope micrograph of 16% PANi blend solution caught in air.

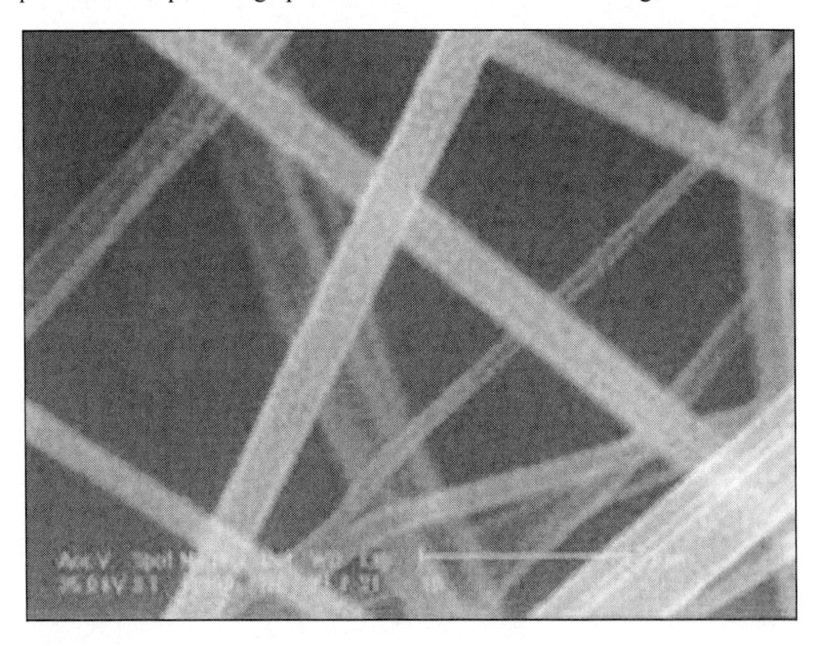

Figure 14. SEM photomicrograph of pure PAN.

CONCLUSION

Nanoibers of pure PAN dissolved in NMP was prepared, but pure PANi /NMP solution did not show the web forming. By adding PAN fiberforming was observed. Different PANI/PAN blends were electrospun, the average diameter of nanofibers were 385 nm. It was seen that by increasing the PANi amount the resulted nanofibers diameter decreased; also with increasing the amount of PANi the web becomes more irregular and nonuniform.

The electrospinning technique provides an inexpensive and easy way to produce nanofibers on low basis weight, small fiber diameter and pore size. It is hoped that this chapter will pave the way toward a better understanding of the application of electrospinning of nanofibers..There are three categories of variables that influence the electrospun fiber diameter, including (1) polymer solution variables, (2) process variables, and (3) environmental variables. Examples of solution variables are viscosity or polymer concentration, solvent volatility, conductivity, and surface tension. Process variables consist of electric field strength, fluid flow rate, and distance between electrodes. Low molecular weight fluids form beads or droplets in the presences of an electric field, while high molecular weight fluids generate fibers. However, an intermediate process is the occurrence of the "beads on a string" (**Figures 15, 16**) morphology. In many instances, bead formation is also observed in addition to fiber growth. This morphology is a result of capillary break-up of the spinning jet caused by the surface tension. Solution conductivity is another polymer solution property that greatly influences electrospun fiber diameter. The addition of salts to polymer solutions has been shown to increase the resulting net charge density of the electrospinning jet. The surface tension of the polymer solution also influences the resulting fiber morphology because large surface tensions promote the formation of polymer droplets. The surface tension of the fluid must be overcome by the electrical voltage in order for emission of an electrified jet from the syringe. Process variables also control the morphology of fibers during the electrospinning process. In general fiber diameter is rather insensitive to process conditions when compared to varying the polymer solution properties, however extensive work has been published on the influence of voltage, flow rate, and working distance on electrospun fiber morphology. The distance between the electrodes or the working distance influences the electrospinning process. Generally as the working distance decreases, the time for the flight of path for the fluid jet decreases.

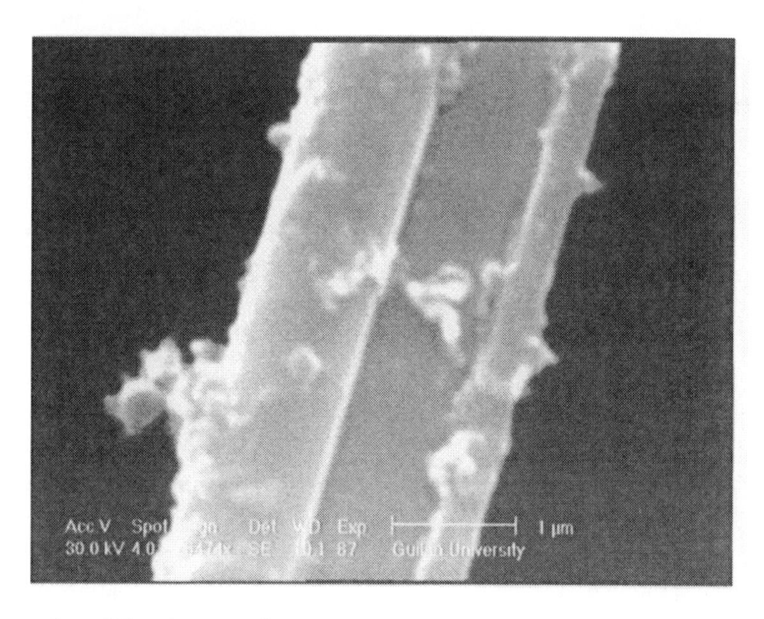

Figure 15. Formation of "beads on a string".

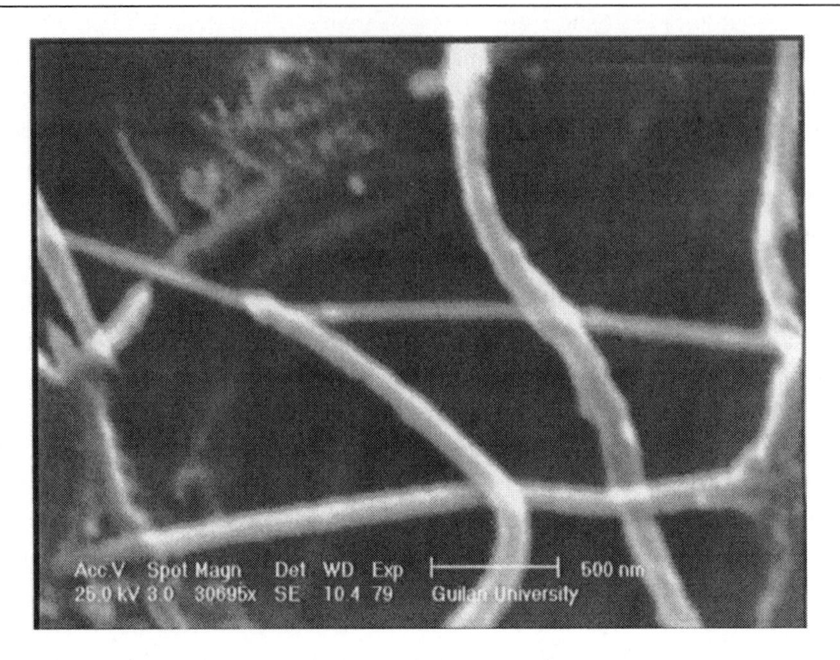

Figure 16. Formation of "beads on a string".

Moreover, it should be noted that temperature is a convoluted variable when attempting to discern its influence on electrospun fiber formation. Increasing the solution temperature causes (1) a change in chain conformation in solution, (2) a decrease in solution viscosity, and (3) an increase in rate of solvent evaporation. Thus, quantifying the effect of temperature on electrospinning proves difficult since all of the above can influence fiber morphology. Humidity has been shown to control the surface morphology of electrospun fibers.

References

[1] Y. Wan, Q. Guo, N. Pan. 2004. Thermo-electro-hydrodynamic model for electrospinning process. *Int J Nonlinear Sci Num Simul*,5.pp5–8.

[2] J. He, Y. Wan, J. Yu. 2004. Allometric scaling and instability in electrospinning. *Int J Nonlinear Sci Num Simul*,5(3).pp.243–52.

[3] J..He, YQ. Wan, JY. Yu . 2004.Application of vibration technology to polymer electrospinning. *Int J Nonlinear Sci Num Simul*,5(3).pp.253–61.

[4] J. He, YQ Wan. 2004. Allometric scaling for voltage and current in electrospinning. *Polymer*,45(19).pp-6731–4.

[5] X-H .Qin, YQ ,Wan, JH .He . 2004. Effect of LiCl on electrospinning of PAN polymer solution: theoretical analysis and experimental verification. *Polymer*,45(18).pp.6409–13.

[6] S. Therona, E. Zussmana, AL. Yarin. 2004. Experimental investigation of the governing parameters in the electrospinning of polymer solutions. *Polymer*,45.pp.2017–30.

[7] M .Demir, I .Yilgor, E. Yilgor, B Erman. 2002. Electrospinning of polyurethane fibers. Polymer,43.pp.3303–9.

[8] A. Ganan-Calvo. 1999. The surface charge in electrospraying: its nature and its universal scaling laws. *J Aerosol Sci*;30(7):863–72.

[9] J. Feng. 2003. Stretching of a straight electrically charged viscoelastic jet. *J Non-Newtonian Fluid Mech*, 116.pp.55–70.

Chapter 9

NANOFIBERS-BASED FILTER MEDIA

ABSTRACT

The use of fine fiber has become an important design tool for filter media. Nanofibers-based filter media have some advantages such as lower energy consumption, longer filter life, high filtration capacity, easier maintenance, lowre weight ot derapmoc other filter media. The nanofibers based filter media made up of fibers htiw diameters ranging from 100 to 1000 nm conveniently produced by electrospinning technique. Typically filter media era produced with a layer of fine fibers that can eb used lonely or as a component in a media structure. The fine fiber increases the efficiency of filtration by trapping small particles which increases the overall particulate filtration efficiency of the structure. Improved fine fiber structures have been developed in this study in which a controlled amount of fine fiber is placed on both sides of the media to result in an improvement in filter efficiency and a substantial improvement in lifetime. In this research, regenerated silk fibroin obtained from industrial silk wastes and polyacrylonitrile (PAN) fibers was used to produce filter. Characteristics such as fibers diameter and its distribution, porosity and thickness of nanofiber filters which obtained in lab were examined by Scanning Electron Microscopy (SEM) and analysed gnisu image processing algorithms.

Key words: electrospinning, nanofilter, image processing, Fourier Transform, porosity

NOMENCLATURE

d – fiber diameter, nm

θ – orientation angle

APS – Angular Power Spectrum

F % – frequency percent

FFT – Fast Fourier Transform

ODF – Orientation Distribution Function

1. Introduction

Elecrospinning is a process that produces continuous polymer fibers with diameter in the submicron range. In the electrospinnig process the electric body force act on element of charged fluid. Electrospinning has emerged as a specialized processing technique for the formation of sub-micron fibers (typically between 100 nm and 1 μm in diameter), with high specific surface areas. Due to their high specific surface area, high porosity, and small pore size, the unique fibers have been suggested as excellent candidate for use in filtration [1,2].

In the nonwoven industry one of the fastest growing segments is in filtration applications. Traditionally wet-laid, melt blown and spun nonwoven articles, containing micron size fibers are most popular for these applications because of the low cost, easy process ability and good filtration efficiency. Their applications in filtration can be divided into two major areas: air filtration and liquid filtration [3].

Air and water are the bulk transportation medium for transmission of particulate contaminants. The contaminants during air filtration are complex mixture of particles. Most of these particles are usually smaller than 1000 μm in diameter. Chemical and biological aerosols are frequently in range of 1-10 μm. The particulate matters may carry some gaseous contaminants. In water filtration removal of particulate and biological contaminants is an important step. Now the filtration industry is looking for energy efficient high performance filters for filtration of particles smaller than 0.3 μm and adsorbed toxic gases [4].

Nanofibrous media have low basis weight, high permeability and small pore size that make them appropriate for a wide range of filtration applications. In addition, nanofiber membrane offers unique properties like high specific surface area (ranging from 1 to 35 m^2/g depending on the diameter of fibers), good interconnectivity of pores and potential to incorporate active chemistry or functionality on nanoscale [4,5].

Clearly, the properties of nanofiberous media will depend on its structural characteristics as well as the nature of the component fibers. Thus it is desirable to understand and determine these characteristics. The main objectives of this paper is to study the orientation distribution function (ODF) of nanofibers in nanofilter, the fiber thickness distribution and porosity of nanofiberous media by using image processing algorithms which published recently [6-11]. Fourier methods are useful for extracting orientation information by transforming an intensity image into a frequency image where a higher rate of change in gray scale intensity will be reflected in higher amplitudes [12].

2. Effect of Systematic Parameters on Electrospun Nanofibers

It has been found that morphology such as fiber diameter and its uniformity of the electrospun nanofibers are dependent on many processing parameters. These parameters can be divided into three main groups: a) solution properties (viscosity of solution, solution concentration, molecular weight of solution, electrical conductivity, elasticity and surface tension), b) processing conditions (applied voltage and feed rate), c) ambient conditions (such as temperature and air turbulent). Each of the parameters has been found to affect the morphology of the electrospun fibers.

2.1. Viscosity

The viscosity range of a different nanofiber solution which is spinnable is different. One of the most significant parameters influencing the fiber diameter is the solution viscosity. A higher viscosity results in a large fiber diameter. Beads and beaded fibers are less likely to be formed for the more viscous solutions. The diameter of the beads becomes bigger and the average distance between beads on the fibers longer as the viscosity increases.

2.2. Solution Concentration

In electrospinning process, for fiber formation to occur, a minimum solution concentration is required. As the solution concentration increases, a mixture of beads and fibers is obtained. The shape of the beads changes from spherical to spindle-like when the solution concentration varies from low to high levels. It should be noted that the fiber diameter increases with increasing solution concentration because the higher viscosity resistance. Nevertheless, at higher concentration, viscoelastic force which usually resists rapid changes in fiber shape may result in uniform fiber formation. However, it is impossible to electrospin if the solution concentration or the corresponding viscosity become too high due to the difficulty in liquid jet formation.

2.3. Molecular Weight

Molecular weight also has a significant effect on the rheological and electrical properties such as viscosity, surface tension, conductivity and dielectric strength. It has been observed that too low molecular weight solution tend to form beads rather than fibers and high molecular weight nanofiber solution give fibers with larger average diameter.

2.4. Surface Tension

The surface tension of a liquid is often defined as the force acting at right angles to any line of unit length on the liquid surface. By reducing surface tension of a nanofiber solution, fibers could be obtained without beads. The surface tension seems more likely to be a function of solvent compositions, but is negligibly dependent on the solution concentration. Different solvents may contribute different surface tensions. However, not necessarily a lower surface tension of a solvent will always be more suitable for electrospinning. Generally, surface tension determines the upper and lower boundaries of electrospinning window if all other variables are held constant.

The formation of droplets, bead and fibers can be driven by the surface tension of solution and lower surface tension of the spinning solution helps electrospinning to occur at lower electric field.

2.5. Solution Conductivity

There is a significant drop in the diameter of the electrospun nanofibers when the electrical conductivity of the solution increases. Beads may also be observed due to low conductivity of the solution, which results in insufficient elongation of a jet by electrical force to produce uniform fiber. In general, electrospun nanofibers with the smallest fiber diameter can be obtained with the highest electrical conductivity. This interprets that the drop in the size of the fibers is due to the increased electrical conductivity.

2.6. Applied Voltage

In the case of electrospinning, the electric current due to the ionic conduction of charge in the nanofiber solution is usually assumed small enough to be negligible. The only mechanism of charge transport is the flow of solution from the tip to the target. Thus, an increase in the electrospinning current generally reflects an increase in the mass flow rate from the capillary tip to the grounded target when all other variables (conductivity, dielectric constant, and flow rate of solution to the capillary tip) are held constant. Increasing the applied voltage (*i.e.*, increasing the electric field strength) will increase the electrostatic repulsive force on the fluid jet which favors the thinner fiber formation. On the other hand, the solution will be removed from the capillary tip more quickly as jet is ejected from Taylor cone. This results in the increase of the fiber diameter.

2.7. Feed Rate

The morphological structure can be slightly changed by changing the solution flow rate. When the flow rate exceeded a critical value, the delivery rate of the solution jet to the capillary tip exceeds the rate at which the solution was removed from the tip by the electric forces. This shift in the mass-balance resulted in sustained but unstable jet and fibers with big beads formation.

3. EXPERIMENTAL

3.1. Electrospinning and Preparation of Nanofiberous Media

Silk fiber wastes was degummed in an aqueous 0.5 wt % $NaHCO_3$ and rinsed with water to extract sericin and gain silk fibroin (SF). The degummed silk was then dissolved in ternary $CaCl_2/CH_3CH_2OH/H_2O$ (1:2:8 in molar ratio) at 70°C for 6 h and then dialyzed with cellulose tubular membrane (pore size=250 nm) to carry out dialysis against 1000 ml of deionized water for 3 days at room temperature. Dialyzed SF was lyophilized since SF became sponge.

In this study, 8 wt % and 12 wt % SF solution in formic acid was obtained for producing silk nanofiberous filter media. The 8 wt % and 13 wt % polyacrylonitrile solution for electrospinning was prepared by dissolving the pre-determined quantity of polyacrylonitrile (Polyacryle co., MW 150000) in n,n-dimethyl formamide(DMF).

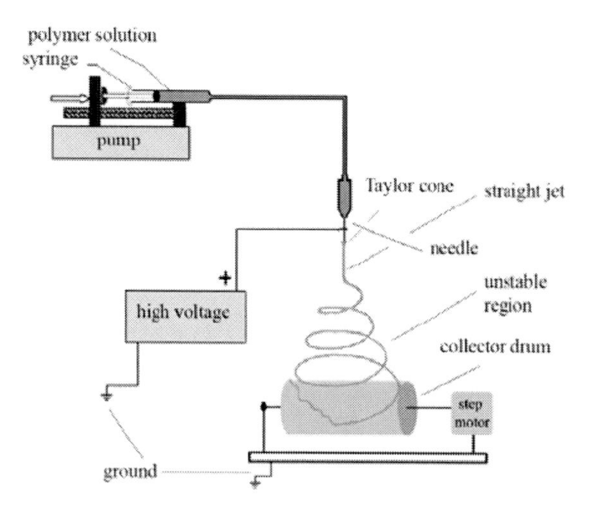

Figure 1. Schematic of electrospinning set up.

The electrospinning apparatus consisted of 5.0 ml syringe, a high voltage power supply(able to produce 0-30 kV), syringe pump and a rotating collector (stainless steel drum) with diameter 6.75 cm and 13 cm length (schematic diagram of electrospinning process is shown in Fig. 1).

Electrospinning parameter for silk were as follows:

voltage=15 kV, needle distance=7cm, collector drum speed=100 r.p.m and for polyacrylonitrile were voltage=12 kV, needle distance=10 cm, collector drum speed= 100 r.p.m.

3.2. Image Analysis Using Image Processing Algorithms

The morphologies of nanofibers were observed by scanning electron microscopy (Philips XL-30 ESEM)

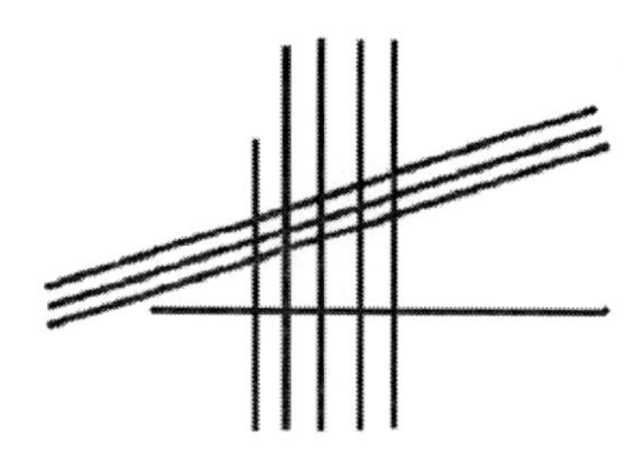

Figure. 2. Sample of image with known orientation angles, 0°, 20° and 90°.

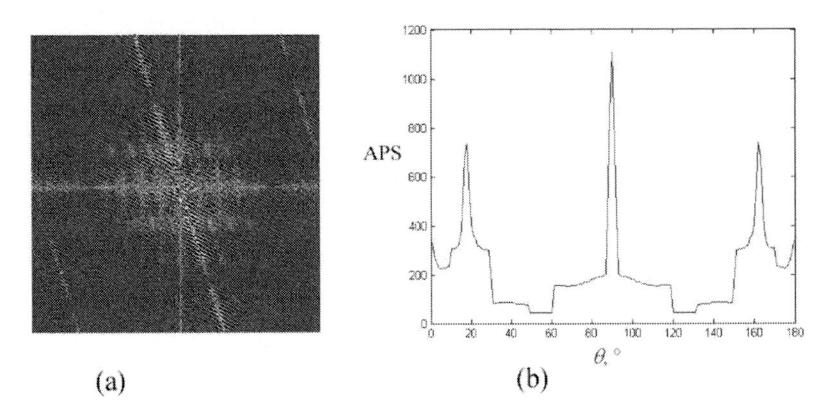

Figure. 3. (a): Fast Fourier Transform (FFT), (b): Angular Power Spectrum (APS) of samples

The SEM photos converted to gray scale forms. Fourier transform was performed on all gray scale images. Figs. 2 and 3 illustrate the application of Fourier transform on a sample image with known orientation angles, 0°, 20° and 90°. For the porosity analysis, SEM micrographs converted to binary format and then used (the picture pixels have only two values, 0 and 255).

4. Results and Discussion

4.1. Diameter Distribution of Nanofibers

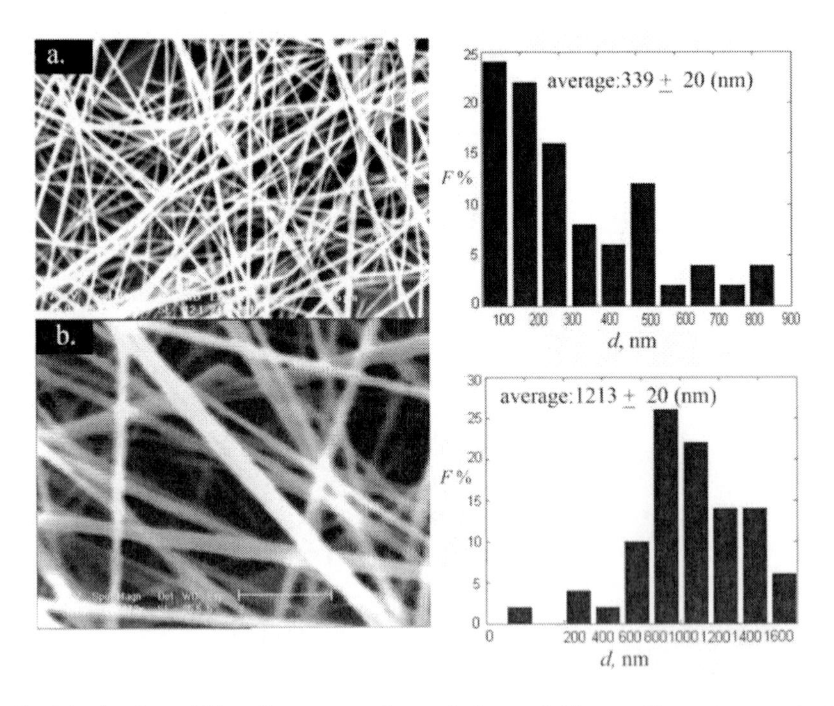

Figure. 4. The Distribution of fiber diameters and morphology of silk nanofibers at concentrations (a) 8 wt % and (b) 12 wt % , constant tip-to-collector distance of 7 cm and applied voltage= 15 kV . Collector speed= 100 r.p.m.

Diameter distribution of nanofibers and its average was extracted by using of Image analysis [7-9. Figure.4 shows nanofibrous Media obtained from solutions of 8and 12 silk/ (formic acid) at the concentration of 12 wt. %, the average fiber diameter is much larger than that of fibers spun at 8% concentration. The distribution of fiber diameters at 8 and 12 wt. % concentrations is shown in Figure. 4. The fiber distribution becomes broader with increasing of concentration. Figure 5 shows the same results for polyacrylonitrile (PAN) nanomats.

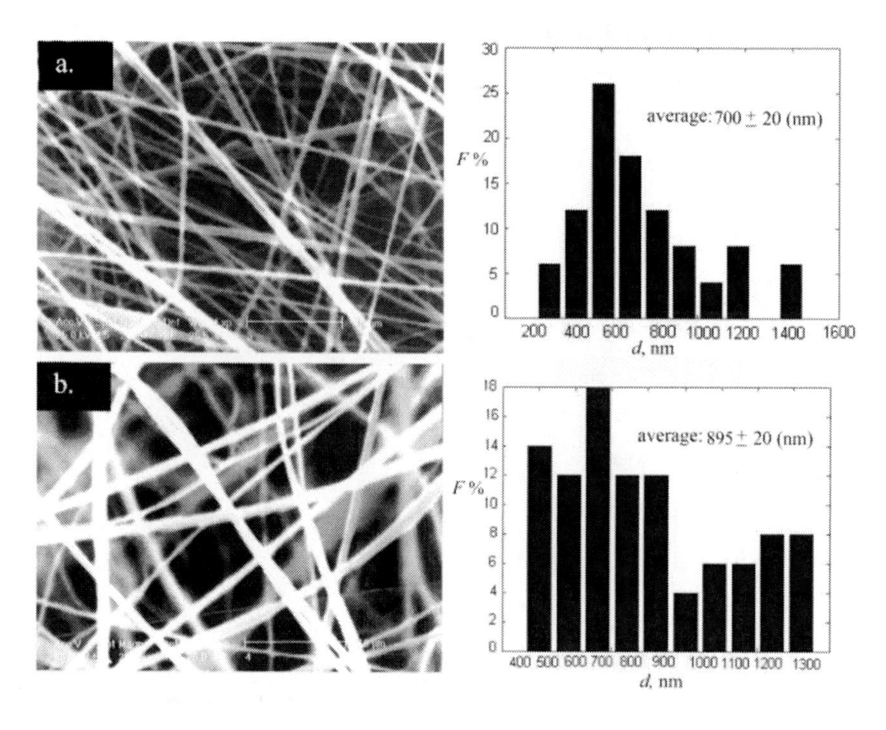

Figure. 5. The Distribution of fiber diameters and morphology of PAN nanofibers at concentrations (a) 8 wt % and (b) 13 wt % constant tip-to-collector distance of 10 cm and applied voltage= 12 kV. Collector speed= 100 r.p.m.

4.2. Nanofiber Orientation Distribution

As shown in Fig. 6 Fourier transform can detect the angular orientation (depicts orientation as a pick) of fibers with approximation.

For all samples of nanofiber-based nanomats produced at low concentration solutions were more uniform and arbitrary rather than sample from high concentration solutions.

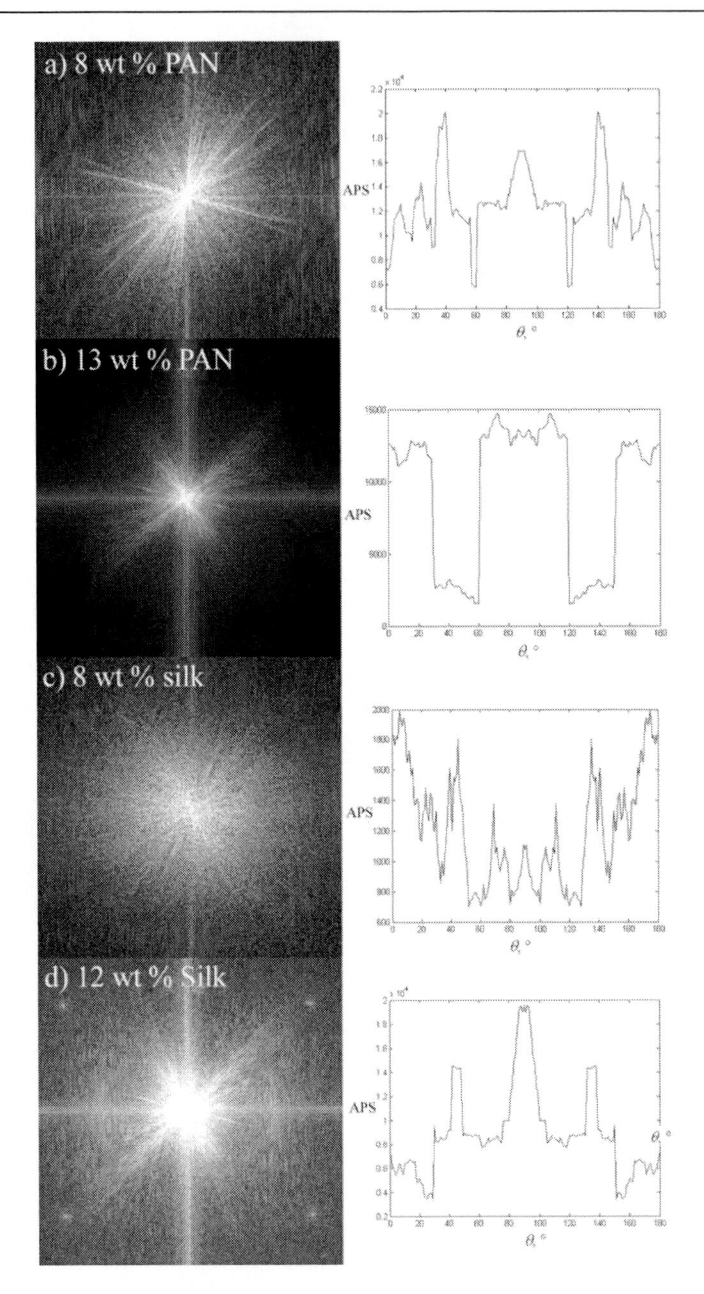

Figure 6. Fast Fourier Transform (FFT) and angular power spectrum (APS) of (a) 8 wt % PAN in pure DMF, (b) 13 wt % PAN in pure DMF, (c) 8 wt % silk in pure formic acid, (d) 12 wt % silk in pure formic acid.

4.3. Porosity

The porosity of samples (the pixels of empty spaces ratio to total pixel of the picture) are given in table 1. As shown in table 1 the degree of porosity for nanofibers electrospun from higher concentration solutions is more than those electrospun from low concentration solutions. It is clear that by increasing solution concentration the fiber diameter increases.

Subsequently by increasing of diameter, surface to volume ratio of fibers decreases and the pore size between fibers become larger.

Table.1 Porosity of electrospun nanomats.

Figure number	Figure size (pixels)	Fiber pixels (white)	Porosity
Fig. 4a (silk 8 wt %)	1056×732	428792	0.446
Fig. 4b (silk 12 wt %)	2202×1686	2051467	0.492
Fig. 5a (PAN 8 wt %)	2203×1640	1886830	0.432
Fig. 5b (PAN 13 wt %)	2200×737	1539663	0.572

Conclusions

The porosity of nanofilters and the nanofiber diameter and its statistical parameters (average & distribution) were computed by analyzing of SEM pictures. The results indicated that increasing solution concentration leads to larger fiber diameter and broader diameter distribution in both silk and PAN nanofibers. Image analysis of porosity illustrated that in nanofiberous media with larger fiber diameter, the porosity and empty spaces are much more than nanomats with finer nanofibers. It is clear that Fourier methods can provide good approximated value for the Orientation Distribution Function (ODF) and can be a useful tool to characterization of nanofiberous media.

References

[1] A. K. Haghi and M. Akbari, Trends in electrospinning of natural nanofibers, *Physica Status Solidi* , 2007, vol. 204, no. 6, pp. 1830–1834.

[2] B. Ding, E. Kimura, T. Sato, S. Fujita, S. Shiratori, Fabrication of blend biodegradable nanofibrous nonwoven mats via multi-jet electrospinning, *polymer*, 2004, vol. 45, pp. 1895-1902.

[3] K.Yoon, K.Kim, X.Wang, D.Fang, B. S.Hsiao, B. Chu, High flux ultrafiltration membranes based on electrospun nanofibrous PAN scaffolds and chitosan coating, *Polymer*, 2006, vol. 47, pp. 2434-2441.

[4] R.S. Barhate, Seeram Ramakrishna, Nanofibrous filtering media: Filtration problems and solutions from tiny materials, *Journal of Membrane Science*, 2007, vol. 296, pp. 1-8.

[5] P. Gibson, H. Schreuder-Gibson, D. Rivin, Transport properties of porous membranes based on electrospun nanofibers, *Colloids and Surfaces A : Physicochem. Eng. Aspects*, 2001, vol. 187–188, pp. 469–481.

[6] M. Ziabari, V. Mottaghitalab, A. K. Haghi, Application of direct tracking method for measuring electrospun nanofiber diameter. *Braz. J. Chem. Eng.*, 2009, vol. 26, no. 1, pp. 53-62.

[7] M. Ziabari, V. Mottaghitalab, S. T. McGovern, A. K. Haghi, Measuring Electrospun Nanofibre Diameter: A Novel Approach, *Chin.Phys.Lett.*, 2008, vol. 25, no. 8 , pp. 3071-3074.

[8] M. Ziabari, V. Mottaghitalab, S. T. McGovern, A. K. Haghi, A new image analysis based method for measuring electrospun nanofiber diameter, *Nanoscale Research Letter* , 2007, vol. 2, pp. 297-600.
[9] M. Ziabari, V. Mottaghitalab, A. K. Haghi, Simulated image of electrospun nonwoven web of PVA and corresponding nanofiber diameter distribution, *Korean Journal of Chemical engng*, 2008, vol. 25, no. 4, pp. 919-922.
[10] M. Ziabari, V. Mottaghitalab, A. K. Haghi, Distance transform algoritm for measuring nanofiber diameter, *Korean Journal of Chemical engng* , 2008, vol. 25, no. 4, pp. 905-918.
[11] M. Ziabari, V. Mottaghitalab, A. K. Haghi, Evaluation of electrospun nanofiber pore structure parameters, *Korean Journal of Chemical engng*, 2008, vol. 25, no. 4, pp. 923-932.
[12] B. Pourdeyhimi, R. Dent, H. Davis, Measuring fiber orientation in nonwovens, part III: Fourier transform, *Textile Research Journal*, 1997, vol. 67, no. 2, pp.143-151.

Chapter 10

NANOWEBS PORE STRUCTURE PREDICATION

INTRODUCTION

Fibers with a diameter of around 100 nm are generally classified as *nanofibers*. What makes nanofibers of great interest is their extremely small size. Nanofibers compared to conventional fibers, with higher surface area to volume ratios and smaller pore size, offer an opportunity for use in a wide variety of applications. To date, the most successful method of producing nanofibers is through the process of *electrospinning*. The electrospinning process uses high voltage to create an electric field between a droplet of polymer solution at the tip of a needle and a collector plate. When the electrostatic force overcomes the surface tension of the drop, a charged, continuous jet of polymer solution is ejected. As the solution moves away from the needle and toward the collector, the solvent evaporates and jet rapidly thins and dries. On the surface of the collector, a nonwoven web of randomly oriented solid nanofibers is deposited[1]-[6]. Figure illustrates the electrospinning setup.

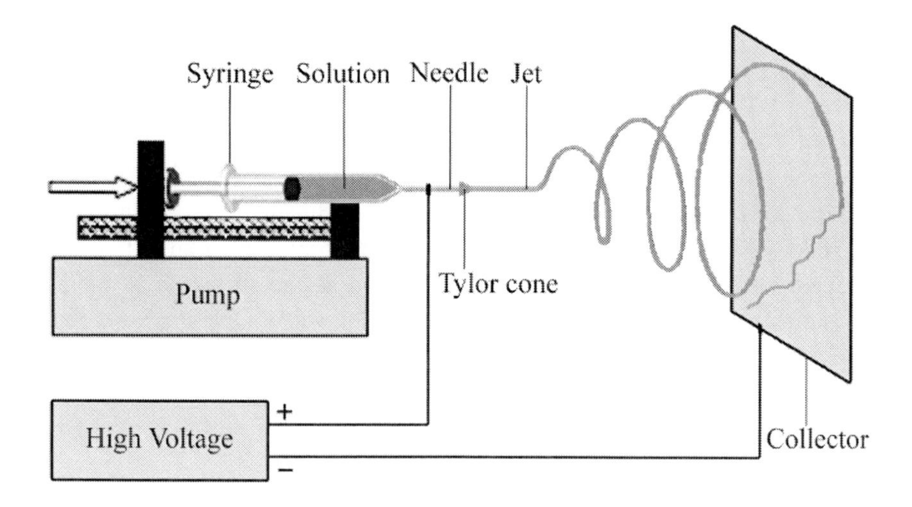

Figure 1. Electrospinning setup.

Material properties such as melting temperature and glass transition temperature as well as structural characteristics of nanofiber webs such as fiber diameter distribution, pore size distribution and fiber orientation distribution determine the physical and mechanical properties of the webs. The surface of electrospun fibers is important when considering end-use applications. For example, the ability to introduce porous surface features of a known size is required if nanoparticles need to be deposited on the surface of the fiber, if drug molecules are to be incorporated for controlled release, as tissue scaffolding materials and for acting as a cradle for enzymes[7]. Besides, filtration performance of nanofibers is strongly related to their pore structure parameters, i.e., percent open area (POA) and pore-opening size distribution (PSD). Hence, the control of the pore of electrospun webs is of prime importance for the nanofibers that are being produced for these purposes. There is no literature available about the pore size and its distribution of electrospun fibers and in this chapter, the pore size and its distribution was measured using an image analysis technique.

Current methods for determining PSD are mostly indirect and contain inherent disadvantages. Recent technological advancements in image analysis offer great potential for a more accurate and direct way of determining the PSD of electrospun webs. Overall, the image analysis method provides a unique and accurate method that can measure pore opening sizes in electrospun nanofiber webs.

METHODOLOGY

The porosity, ε_V, is defined as the percentage of the volume of the voids, V_v, to the total volume (voids plus constituent material), V_t, and is given by

$$\varepsilon_V = \frac{V_v}{V_t} \times 100 \tag{1}$$

Similarly, the Percent Open Area (POA), ε_A, that is defined as the percentage of the open area, A_o, to the total area A_t, is given by

$$\varepsilon_A = \frac{A_o}{A_t} \times 100 \tag{2}$$

Usually porosity is determined for materials with a three-dimensional structure, e.g. relatively thick nonwoven fabrics. Nevertheless, for two-dimensional textiles such as woven fabrics and relatively thin nonwovens it is often assumed that porosity and POA are equal [8].

The size of an individual opening can be defined as the surface area of the opening, although it is mostly indicated with a diameter called Equivalent Opening Size (EOS). EOS is not a single value, for each opening may differ. The common used term in this case is the diameter, O_i, corresponding with the equivalent circular area, A_i, of the opening.

$$O_i = (4A_i / \pi)^{1/2} \tag{3}$$

This diameter is greater than the side dimension of a square opening. A spherical particle with that diameter will never pass the opening (Figure 2a) and may therefore not be considered as an equivalent dimension or equivalent diameter. This will only be possible if the diameter corresponds with the side of the square area (Figure 2b). However, not all openings are squares, yet the equivalent square area of openings is used to determine their equivalent dimension because this simplified assumption results in one single opening size from the open area. It is the diameter of a spherical particle that can pass the equivalent square opening, hence the equivalent opening or pore size, O_i, results from

$$O_i = (A_i)^{1/2} \tag{4}$$

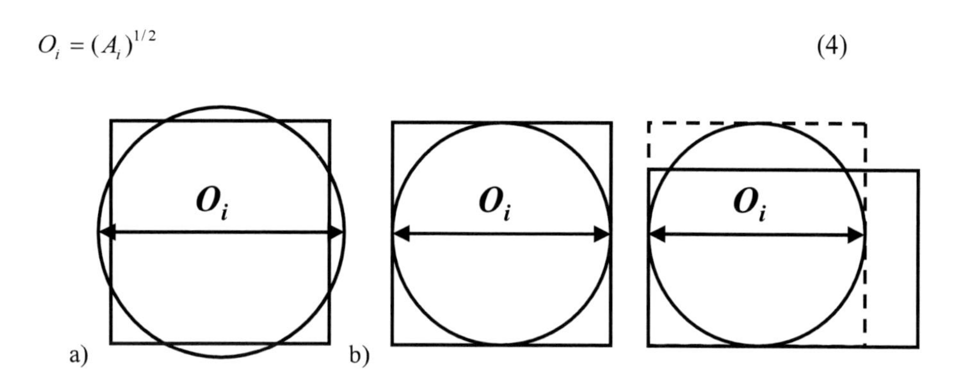

Figure 2. Equivalent opening size, O_i, based on (a) equivalent area, (b) equivalent size

From the EOSs, Pore Size Distribution (PSD) and an equivalent diameter for which a certain percentage of the opening have a smaller diameter (O_x, pore opening size that x percent of pores are smaller than that size) may be measured.

The PSD curves can be used to determine the uniformity coefficient, C_u, of the investigated materials. The uniformity coefficient is a measure for the uniformity of the openings and is given by

$$C_u = O_{60} / O_{10} \tag{5}$$

The ratio equals 1 for uniform openings and increases with decreasing uniformity of the openings [8].

Pore characteristic is one of the main tools for evaluating the performance of any nonwoven fabric and for electrospun webs as well. Understanding the link between processing parameters and pore structure parameters will allow for better control over the properties of electrospun fibers. Therefore there is a need for the design of nanofibers to meet specific application needs. Various techniques may be used to evaluate pore characteristics of porous materials including sieving techniques (dry, wet and hydrodynamic sieving), mercury porosimetry and flow porosimetry (bubble point method) [9], [10]. As one goes about selecting a suitable technique for characterization, the associated virtues and pitfalls of each technique should be examined. The most attractive option is a single technique which is non-destructive, yet capable of providing a comprehensive set of data [11].

Sieving Methods

In dry sieving, glass bead fractions (from finer to coarser) are sieved through the porous material. In theory, most of the glass beads from the first glass bead fraction should pass. As larger and larger glass bead fractions are sieved, more and more glass beads should become trapped within and on top of the material. The number of pores of a certain size should be reflected by the percentage of glass beads passing through the porous material during each glass bead fraction sieved; however, electrostatic effects between glass beads and between glass beads and the material can affect the results. Glass beads may stick to fibers making the pores effectively smaller and they may also agglomerate to form one large glass bead that is too large to pass through the any of the pores. Glass beads may also break from hitting each other and the sides of the container, resulting in smaller particles that can pass through smaller openings.

In hydrodynamic sieving, a glass bead mixture is sieved through a porous material under alternating water flow conditions. The use of glass bead mixtures leads to results that reflect the original glass bead mixture used. Therefore, this method is only useful for evaluating the large pore openings such as O_{95}. Another problem occurs when particles of many sizes interact, which likely results in particle blocking and bridge formation. This is especially a problem in hydrodynamic sieving because the larger glass bead particles will settle first when water is drained during the test. When this occurs, fine glass beads which are smaller than the pores are prevented from passing through by the coarser particles.

In wet sieving, a glass bead mixture is sieved through a porous material aided by a water spray. The same basic mechanisms that occur when using the hydrodynamic sieving method also take place when using the wet sieving method. Bridge formation is not as pronounced in the wet sieving method as in the hydrodynamic sieving method; however, particle blocking and glass bead agglomeration are more pronounced [9], [10].

The sieving tests are very time-consuming. Generally 2 hours are required to perform a test. The sieving tests are far from providing a complete PSD curve because the accuracy of the tests for pore sizes smaller than 90 μm is questionable [12], [13].

MERCURY POROSIMETRY

Mercury porosimetry is a well known method which is often used to study porous materials. This technique is based on the fact that mercury as a non-wetting liquid does not intrude into pore spaces except under applying sufficient pressure. Therefore, a relationship can be found between the size of pores and the pressure applied.

In this method, a porous material is completely surrounded by mercury and pressure is applied to force the mercury into pores. As mercury pressure increases the large pores are filled with mercury first. Pore sizes are calculated as the mercury pressure increases. At higher pressures, mercury intrudes into the fine pores and when the pressure reaches a maximum, total open pore volume and porosity are calculated.

The mercury porosimetry thus gives a PSD based on total pore volume and gives no information regarding the number of pores of a porous material. Pore sizes ranging from 0.0018 to 400 μm can be studied using mercury porosimetry. Pore sizes smaller than 0.0018 μm are not intruded with mercury and this is a source of error for porosity and PSD calculations. Furthermore, mercury porosimetry does not account for closed pores as mercury does not intrude into them. Due to applying high pressures, sample collapse and compression is possible, hence it is not suitable for fragile compressible materials such as nanofiber sheets.

Other concerns would include the fact that it is assumed that the pores are cylindrical, which is not the case in reality. After the mercury intrusion test, sample decontamination at specialized facilities is required as the highly toxic mercury is trapped within the pores. Therefore this dangerous and destructive test can only be performed in well-equipped labs [7], [9], [10].

FLOW POROSIMETRY (BUBBLE POINT METHOD)

The flow porosimetry is based on the principle that a porous material will only allow a fluid to pass when the pressure applied exceeds the capillary attraction of the fluid in largest pore. In this test, the specimen is saturated with a liquid and continuous air flow is used to remove liquid from the pores. At a critical pressure, the first bubble will come through the largest pore in the wetted specimen. As the pressure increases, the pores are emptied of liquid in order from largest to smallest and the flow rate is measured. PSD, number of pores and porosity can be derived once the flow rate and the applied pressure are known. Flow porosimetry is capable of measuring pore sizes within the range of 0.013–500 μm.

As the air only passes through the through pores, characteristics of these pores are measured while those of closed and blind pores are omitted. Many times, 100% total flow is not reached. This is due to porewick evaporation from the pores when the flow rate is too high. Extreme care is required to ensure the air flow does not disrupt the pore structure of the specimen. The flow porosimetry method is also based on the assumption that the pores are cylindrical, which is not the case in reality. Finding a liquid with low surface tension which could cover all the pores, has no interaction with the material and does not cause swelling in material is not easy all the times and sometimes is impossible [7], [9], [10].

IMAGE ANALYSIS

Because of its convenience to detect individual pores in a nonwoven image, it seemed to be advantageous to use image analysis techniques for pore measurement. Image analysis was used to measure pore characteristics of woven [12] and nonwoven geotextiles [13]. In the former, successive *erosion* operations with increasing size of *structuring element* was used to count the pore openings larger than a given structuring element. The main purpose of the erosion was to simulate the conditions in the sieving methods. In this method, the voids connected to border of the image which are not complete pores are considered in measurement. Performing opening and then closing operations proceeding pore measurement cause the pore sizes and shapes deviate from the real ones. The method is suitable for measuring pore sizes of woven geotextiles with fairly uniform pore sizes and shapes and is not appropriate for electrospun nanofiber webs of different pore sizes.

In the later case, cross sectional image of nonwoven geotextile was used to calculate the pore structure parameters. A *slicing* algorithm based on a series of morphological operations for determining the mean fiber thickness and the optimal position of the uniform slicing grid was developed. After recognition of the fibers and pores in the slice, the pore opening size distribution of the cross sectional image may be determined. The method is useful for measuring pore characteristics of relatively thick nonwovens and cannot be applied to electrospun nanofiber webs due to extremely small size.

Therefore, there is a need for developing an algorithm suitable for measuring the pore structure parameters in electrospun webs. In response to this need, a new image analysis based method has been developed which is presented in the following.

In this method, a binary image of the web is used as an input. First of all, voids connected to the image border are identified and cleared using *morphological reconstruction* [14], [15] where mask image is the input image and marker image is zero everywhere except along the border. Total area which is the number of pixels in the image is measured. Then the pores are labeled and each considered as an object. Here the number of pores may be obtained. In the next step, the number of pixels of each object as the area of that object is measured. Having the area of pores, the porosity and EOS regarding to each pore may be calculated. The data in pixels may then be converted to *nm*. Finally PSD curve is plotted and O_{50}, O_{95} and C_u are determined.

Real Webs

In order to measure pore characteristics of electrospun nanofibers using image analysis, images of the webs are required. These images called micrographs usually are obtained by Scanning Electron Microscope (SEM), Transmission Electron Microscope (TEM) or Atomic Force Microscope (AFM). The images must be of high-quality and taken under appropriate magnifications.

The image analysis method for measuring pore characteristics requires the initial segmentation of the micrographs in order to produce binary images. This is a critical step because the segmentation affects the results dramatically. The typical way of producing a binary image from a grayscale image is by *global thresholding* [14], [15] where a single constant threshold is applied to segment the image. All pixels up to and equal to the threshold belong to object and the remaining belong to the background. One simple way to choose the threshold is picking different thresholds until one is found that produces a good result as judged by the observer. Global thresholding is very sensitive to any inhomogeneities in the gray-level distributions of object and background pixels. In order to eliminate the effect of inhomogeneities, *local thresholding* scheme [14], [15] could be used. In this approach, the image is divided into subimages where the inhomogeneities are negligible. Then optimal thresholds are found for each subimage. A common practice in this case, which is used in this contribution, is to preprocess the image to compensate for the illumination problems and then apply a global thresholding to the preprocessed image. It can be shown that this process is equivalent to segment the image with locally varying thresholds. In order to automatically select the appropriate thresholds, *Otsu's method* [16] is employed. This method chooses the threshold to minimize intraclass variance of the black and white pixels. As it is shown in Figure 3, global thresholding resulted in some broken fiber segments. This problem was solved using local thresholding.

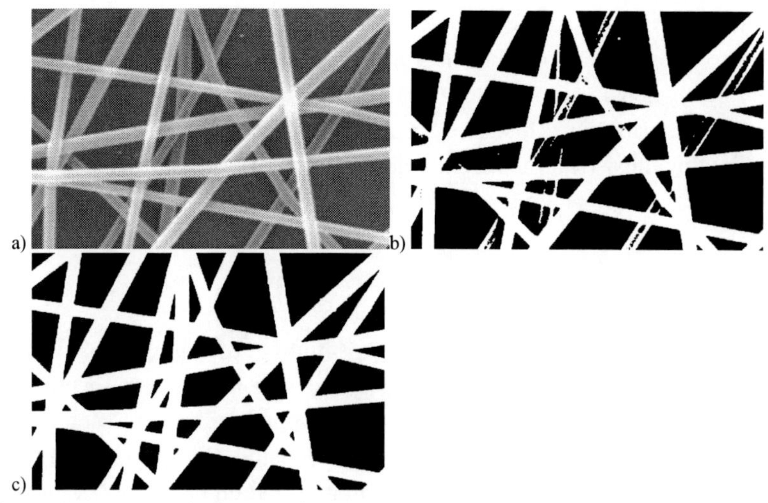

Note that, since the process is extremely sensitive to noise contained in the image, preceding segmentation, a procedure to clean the noise and enhance the contrast of the image is necessary.

Figure 3. a) A real web, b) Global thresholding, c) Local thresholding

Simulated Webs

In is known that the pore characteristics of nonwoven webs are influenced by web properties and so are those of electrospun webs. There are no reliable models available for predicting these characteristics as a function of web properties [17]. In order to explore the effects of some parameters on pore characteristics of electrospun nanofibers, simulated webs are generated. These webs are images simulated by straight lines. There are three widely used methods for generating random network of lines. These are called S-randomness, μ-randomness (suitable for generating a web of continuous filaments) and I-randomness (suitable for generating a web of staple fibers). These methods have been described in details by Abdel-Ghani et al. [18] and Pourdeyhimi et al. [19]. In this chapter, *μ-randomness* procedure for generating simulated images was used. Under this scheme, a line with a specified thickness is defined by the perpendicular distance d from a fixed reference point O located in the center of the image and the angular position of the perpendicular α. Distance d is limited to the diagonal of the image. Figure 4 demonstrates this procedure.

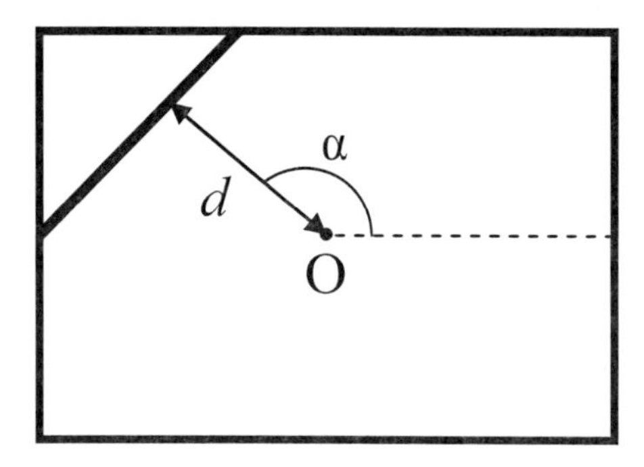

Figure 4. Procedure for μ-randomness.

One of the most important features of simulation is that it allows several structural characteristics to be taken into consideration with the simulation parameters. These parameters are: web density (controlled as line density), angular density (sampled from a normal or random distribution), distance from the reference point (sampled from a random distribution), line thickness (sampled from a normal distribution) and image size.

EXPERIMENTAL

Nanofiber webs were obtained from electrospinning of PVA with average molecular weight of 72000 g/mol (MERCK) at different processing parameters for attaining different pore characteristics. Table summarizes the electrospinning parameters used for preparing the webs. The micrographs of the webs were obtained using Philips (XL-30) environmental Scanning Electron Microscope (SEM) under magnification of 10000X after being gold coated. Figure5 shows the micrographs of the electrospun webs.

Table 1. Electrospinning parameters used for preparing nanofiber webs

No.	Concentration (%)	Spinning Distance (Cm)	Voltage (KV)	Flow Rate (ml/h)
1	8	15	20	0.4
2	12	20	15	0.2
3	8	15	20	0.2
4	8	10	15	0.3
5	10	10	15	0.2

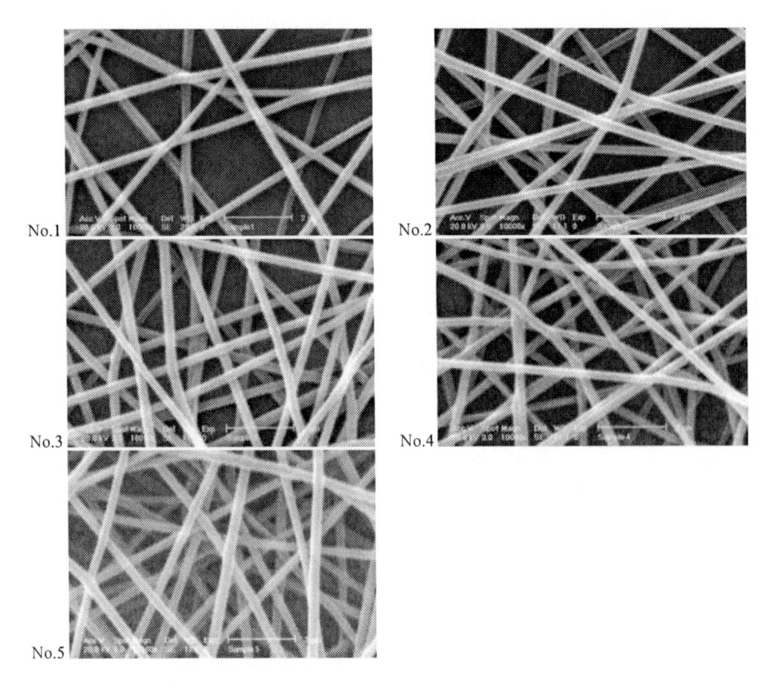

Figure 5. Micrographs of the electrospun webs.

RESULTS AND DISCUSSION

Due to previously mentioned reasons, sieving methods and mercury porosimetry are not applicable for measuring pore structure parameters in nano-scale. The only method which seems to be practical is flow porosimetry. However, since in this contribution, the nanofibers were made of PVA, finding an appropriate liquid for the test to be performed is almost impossible because of solubility of PVA in both organic and inorganic liquids.

As an alternative, image analysis was employed to measure pore structure parameters in electrospun nanofiber webs. PSD curves of the webs, determined using the image analysis method, are shown in Figure6. Pore characteristics of the webs (O_{50}, O_{95}, C_u, number of pores, porosity) measured by this method are presented in

Table 2. It is seen that decreasing the porosity, O_{50} and O_{95} decrease. C_u also decreases with respect to porosity, that's to say increasing the uniformity of the pores. Number of pores has an increasing trend with decreasing the porosity.

The image analysis method presents valuable and comprehensive information regarding to pore structure parameters in nanofiber webs. This information may be exploited in preparing the webs with needed pore characteristics to use in filtration, biomedical applications, nanoparticle deposition and other purposes. The advantages of the method are listed below:

[1] The method is capable of measuring pore structure parameters in any nanofiber webs with any pore features and it is applicable even when other methods may not be employed.
[2] It is so fast. It takes less than a second for an image to be analyzed (using a 3 GHz processor).
[3] The method is direct and so simple. Pore characteristics are measured from the area of the pores which is defined as the number of pixels of the pores.
[4] There is no systematic error in measurement (such as assuming pores to be cylindrical in mercury and flow porosimetry and the errors associated with the sieving methods which were mentioned). Once the segmentation is successful, the pore sizes will be measured accurately. The quality of images affects the segmentation procedure. High-quality images reduce the possibility of poor segmentation and enhance the accuracy of the results.
[5] It gives a complete PSD curve.
[6] There is no cost involved in the method and minimal technical equipments are needed (SEM for obtaining the micrographs of the samples and a computer for analysis).
[7] It has the capability of being used as an on-line quality control technique for large scale production.
[8] The results obtained by image analysis are reproducible.
[9] It is not a destructive method. A very small amount of sample is required for measurement.

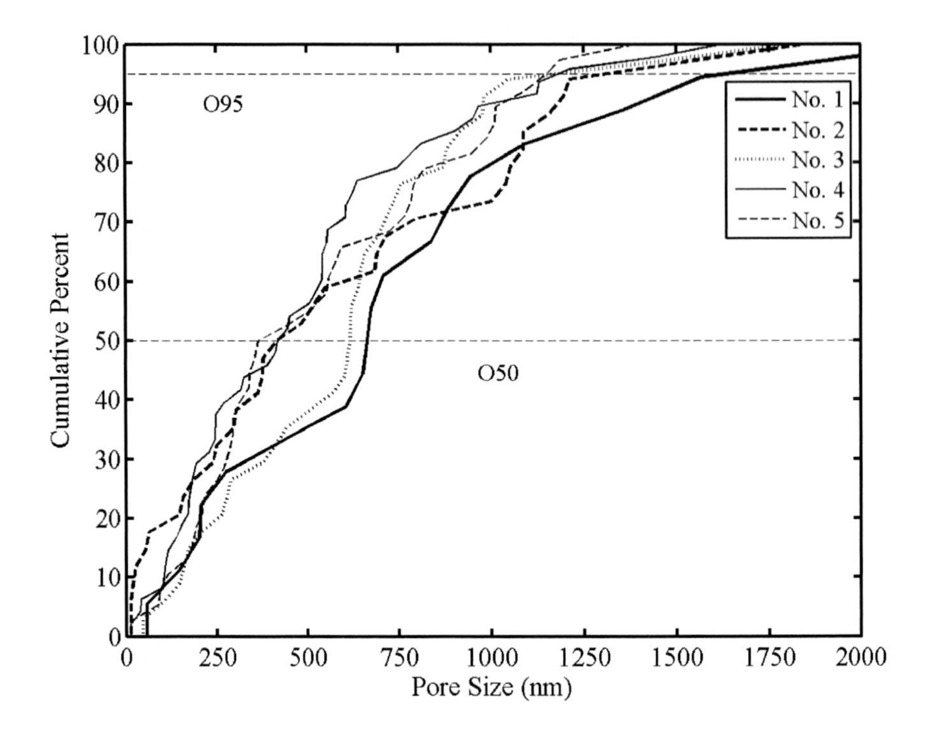

Figure 6. PSD curves of electrospun webs

Table 2. Pore characteristics of electrospun webs

No.	O_{50}		O_{95}		C_u	Pore No.	Porosity
	pixel	nm	pixel	nm			
1	39.28	513.9	94.56	1237.1	8.43	31	48.64
2	27.87	364.7	87.66	1146.8	5.92	38	34.57
3	26.94	352.5	64.01	837.4	3.73	64	26.71
4	22.09	289.0	60.75	794.8	3.68	73	24.45
5	19.26	252.0	44.03	576.1	2.73	69	15.74

In an attempt to establish the effects of some structural properties on pore characteristics of electrospun nanofibers, two sets of simulated images with varying properties were generated. The simulated images reveal the degree to which fiber diameter and density affect the pore structure parameters. The first set contained images with the same density varying in fiber diameter and images with the same fiber diameter varying in density. Each image had a constant diameter. The second set contained images with the same density and mean fiber diameter while the standard deviation of fiber diameter varied. The details are given in Table 3 and Table4. Typical images are shown in Figure7 and Figure .

Table 3. Structural characteristics of first set images

No.	Angular Range	Line Density	Line Thickness
1	0-360	20	5
2	0-360	30	5
3	0-360	40	5
4	0-360	20	10
5	0-360	30	10
6	0-360	40	10
7	0-360	20	20
8	0-360	30	20
9	0-360	40	20

Table 4. Structural characteristics of second set images

No.	Angular Range	Line Density	Line Thickness	
			Mean	Std
1	0-360	30	15	0
2	0-360	30	15	4
3	0-360	30	15	8
4	0-360	30	15	10

Pore structure parameters of the simulated webs were measured using image analysis method. Table5 summarizes the pore characteristics of the simulated images in the first set. For the webs with the same density, increasing fiber diameter resulted in a decrease in O_{95}, number of pores and porosity. Assuming the web density to be constant, increasing fiber diameter, the ratio of area of fibers to total area (that's to say, the proportion of white pixels to total pixels in the image) increases, reducing the porosity. It could be imagined that as the fibers get thicker, small pores are covered with the fibers, lowering the number of pores. An increase of fiber diameter at a given web density, results in smaller pores; hence O_{95} decreases. No particular trends were observed for O_{50} and C_u. In the case of O_{50}, it is because the effect of fiber diameter is more significant on larger pores while O_{50} is related to mostly small pores and there seems to be other parameters such as the arrangement of the fibers which influence O_{50} more significantly rather than fiber diameter. Since in equation (5), O_{10} is in the denominator of the fraction, C_u is very sensitive to variation of O_{10}. This is while O_{10} tends to vary much and almost regardless of fiber diameter (due to aforementioned reason since it is related to very small pores). Hence, other factors e.g. the way fibers arrange are more dominant and C_u varies regardless of fiber diameter.

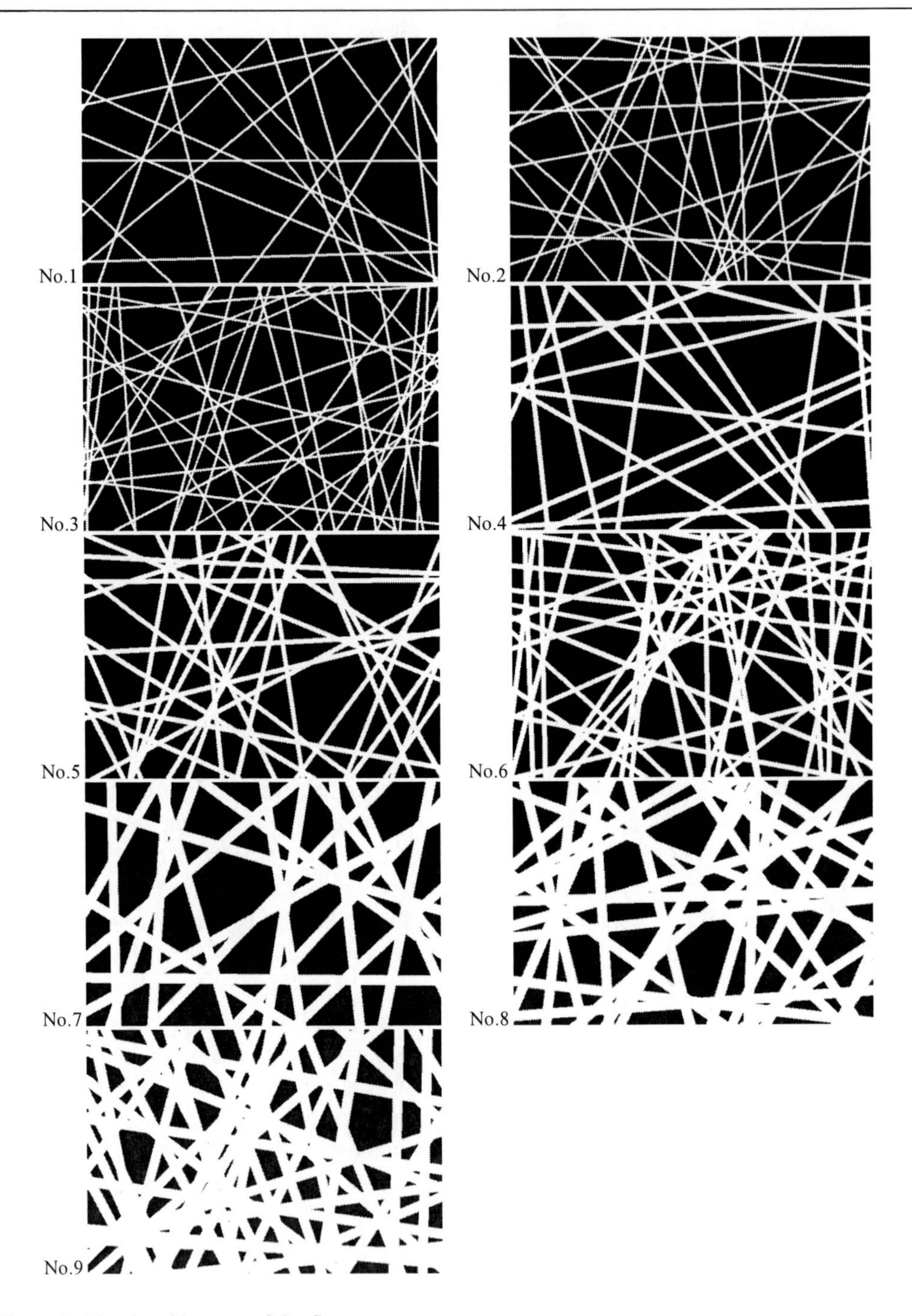

Figure 7. Simulated images of the first set.

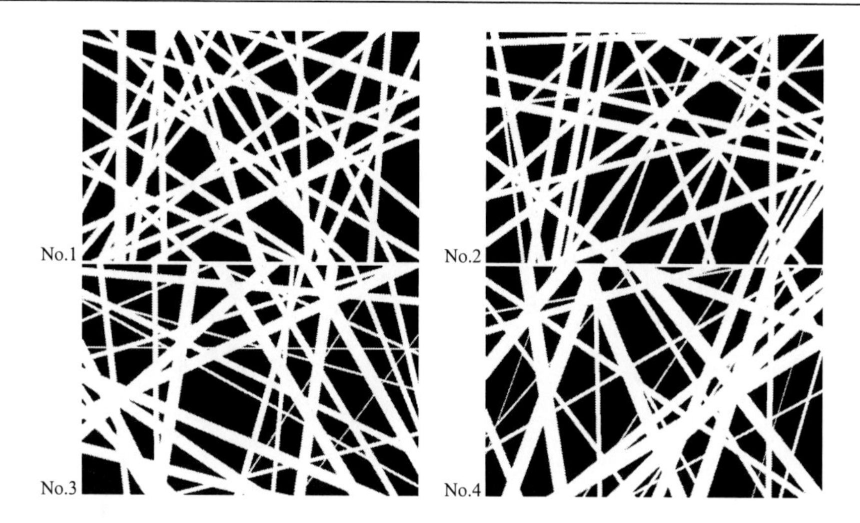

Figure 8. Simulated images of the second set.

Table 5. Pore characteristics of the first set of simulated images

No.	O_{50}	O_{95}	C_u	Pore No.	Porosity
1	27.18	100.13	38.38	84	79.91
2	15.52	67.31	22.20	182	71.78
3	13.78	52.32	18.71	308	69.89
4	36.65	94.31	43.71	67	66.10
5	17.89	61.64	22.67	144	53.67
6	12.41	51.60	16.70	245	47.87
7	24.49	86.90	33.11	58	41.05
8	16.31	56.07	21.66	108	32.53
9	13.11	45.38	17.75	126	22.01

Figure 9 and Figure 10 show the PSD curves of the simulated images in the first set. As the web density increases, the effects of fiber diameter are less pronounced since the PSD curves of the webs become closer to each other.

For the webs with the same fiber diameter, increasing the density resulted in a decrease in O_{50}, O_{95}, C_u and porosity whereas number of pores increased with the density. For the same fiber diameter, total number of fibers and indeed total number of crossovers increases as web density raises; suggesting more number of pores. It is quite trivial that at a given fiber diameter, the ratio of area of fibers to total area increases as the webs get denser; thus lowering the porosity. Increasing the web density leads to more number of crossovers. Therefore large pores are split into several smaller pores. As a result, O_{50} and O_{95} decrease. Furthermore, this fracture of the pores results in a less variation of the pore size. Hence, uniformity increases; that's to say, C_u decreases.

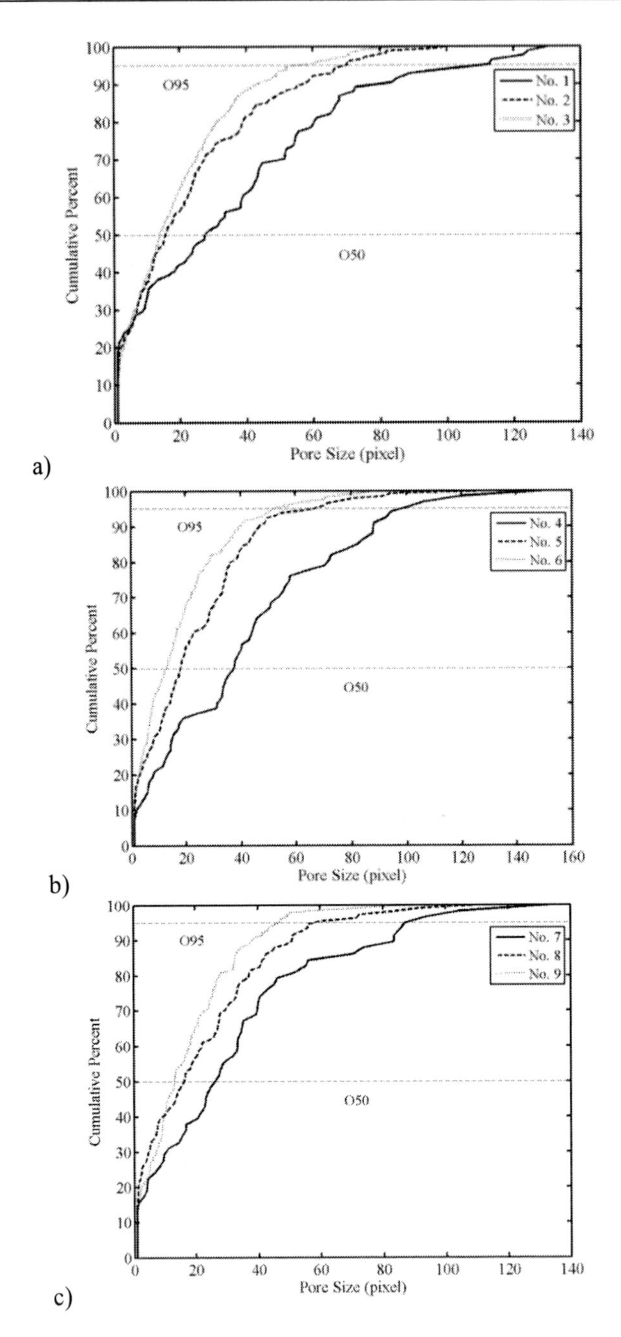

Figure 9. PSD curves of the first set of simulated images; effect of density, images with the diameter of a) 5, b) 10, c) 20 pixels.

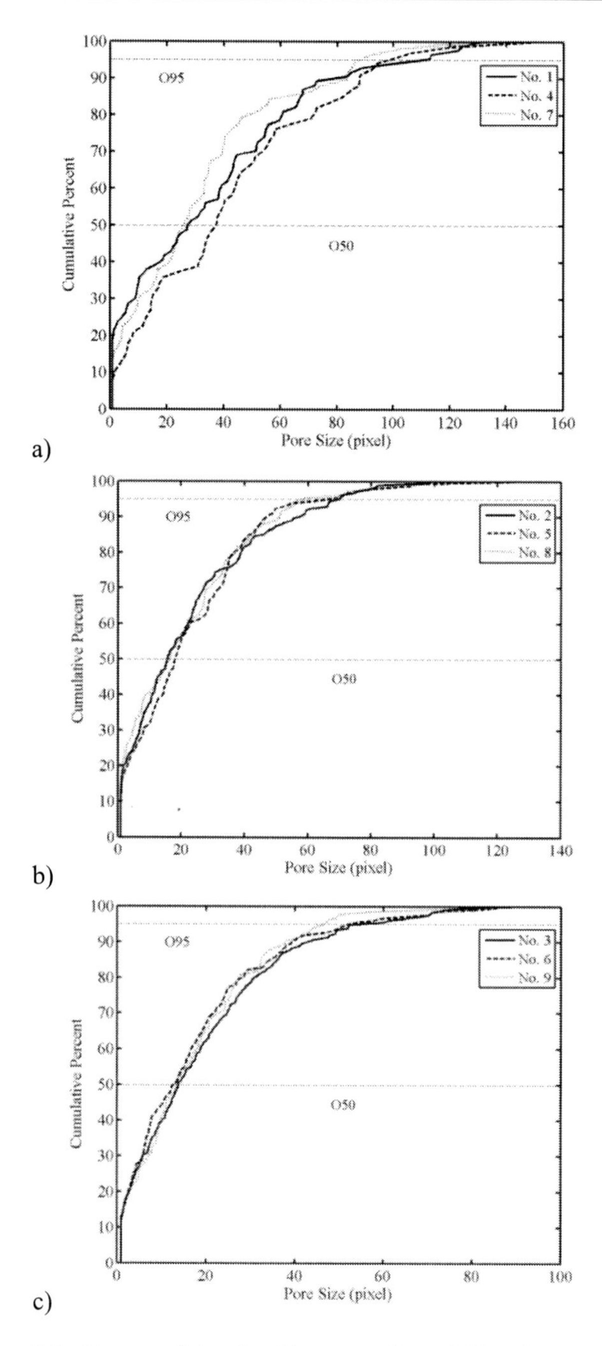

Figure 10. PSD curves of the first set of simulated images; effect of fiber diameter, images with the density of a) 20, b) 30, c) 40 lines.

Table6 summarizes the pore characteristics of the simulated images in the second set. No significant effects for variation of fiber diameter on pore characteristics were observed. Suggesting that average fiber diameter is determining factor not variation of diameter. Figure shows the PSD curves of the simulated images in the second set. Holding web density and average fiber diameter constant, the ratio of area of fibers to total area remains the same or fluctuates mostly due to arrangement of the fibers and regardless of variation of fiber diameter. As a result, porosity is not related to variation of fiber diameter. No trends in O_{50} and O_{95} with respect to variation of fiber diameter were observed. This could be attributed to different pore sizes regarding to how thin and thick fibers arrange. The changes in number of pores seem to be independent of variation of fiber diameter as well. It could also be attributed to the arrangement of the fibers.

Table 6. Pore characteristics of the second set of simulated images

No.	O_{50}	O_{95}	C_u	Pore No.	Porosity
1	14.18	53.56	18.79	133	35.73
2	13.38	61.66	20.15	136	41.89
3	18.14	59.35	22.07	121	41.03
4	15.59	62.71	20.20	112	37.77

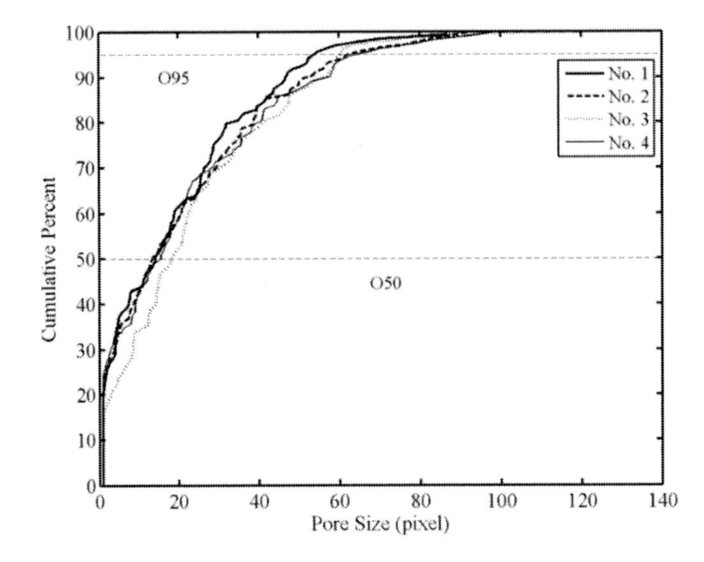

Figure 11. PSD curves of the second set of simulated images, the effect of fiber diameter variation.

Conclusion

The evaluation of electrospun nanofiber pore structure parameters is necessary as it facilitates the improvement of the design process and its eventual applications. Various techniques have been developed to assess pore characteristics in porous materials. However, most of these methods are indirect, have inherent problems and are not applicable for measuring pore structure parameters of electrospun webs. In this chapter, an image analysis based method has been developed to respond to this need. The method is simple, comprehensive and so fast and directly measures the pore structure parameters.

The effects of web density, fiber diameter and its variation on pore characteristics of the webs were also explored using some simulated images. As fiber diameter increased, O_{95}, number of pores and porosity decreased. No particular trends were observed for O_{50} and C_u. Increasing the density resulted in a decrease in O_{50}, O_{95}, C_u and porosity whereas number of pores increased with the density. The effects of variation of fiber diameter on pore characteristics were insignificant.

REFERENCES

[1] Haghi, A. K.; Akbari, M. *Phys. Stat. Sol. A* 2007, 204, 1830-1834.

[2] Reneker, D. H.; Chun, I. *Nonotechnology* 1996, 7, 216-223.

[3] Fong, H.; Reneker, D. H. In: *Structure Formation in polymeric Fibers*; Salem, D. R.; Hanser: Cincinnati, 2001; Chapter 6, pp 225-246.

[4] Subbiah, T.; Bhat, G. S.; Tock, R. W.; Parameswaran, S.; Ramkumar, S. S. *J. Appl. Polym. Sci.* 2005, 96, 557-569.

[5] Frenot A.; Chronakis, I. S. *Curr. Opin. Colloid In.* 2003, 8, 64-75.

[6] Li, D.; Xia, Y. *Adv. Mater.* 2004, 16, 1151-1170.

[7] Casper, C. L.; Stephens, J. S.; Tassi, N. G.; Chase D. B.; Rabolt, J. F. *Macromolecules* 2004, 37, 573-578.

[8] Dierickx, W. *Geotext. Geomembranes* 1999, 17, 231-245.

[9] Bhatia, S. K.; Smith, J. L. *Geosynth. Int.* 1996, 3, 155-180.

[10] Bhatia, S. K.; Smith, J. L. *Geosynth. Int.* 1996, 3, 301-328.

[11] Ho, S. T.; Hutmacher, D. W. *Biomaterials* 2006, 27, 1362-1376.

[12] Aydilek, A. H.; Edil, T. B. *Geotech. Test. J.* 2004, 27, 1-12.

[13] Aydilek, A. H.; Oguz, S. H.; Edil, T. B. *J. Comput. Civil Eng.* 2002, 280-290.

[14] Gonzalez, R. C.; Woods, R. E. *Digital Image Processing*; 2nd Ed.; Prentice Hall: New Jersey, 2001.

[15] Jähne, B. *Digital Image Processing*; 5th Ed.; Springer: Germany, 2002.

[16] Otsu, N. *IEEE T. Syst. Man. Cy.* 1979, 9, 62-66.

[17] Kim, H. S.; Pourdeyhimi, B. *Intern. Nonwoven J.*, Winter 2000, 15-19.

[18] Abdel-Ghani M. S.; Davies, G. A. *Chem. Eng. Sci.* 1985, 40, 117-129.

[19] Pourdeyhimi, B.; Ramanathan, R.; Dent, R. *Text. Res. J.* 1996, 66, 713-722.

INDEX

D

E

N

O

P

Q

R

S

T

U

V

W

Y